CIRIA Special Publication 97   1994

ENVIRONMENTAL HANDBOOK FOR BUILDING AND CIVIL ENGINEERING PROJECTS

DESIGN AND SPECIFICATION

CONSTRUCTION INDUSTRY RESEARCH
AND INFORMATION ASSOCIATION
6 Storey's Gate
Westminster
London SW1P 3AU
Tel 071-222 8891
Fax 071-222 1708

THOMAS TELFORD SERVICES LTD
Thomas Telford House
1 Heron Quay
London E14 4JD
Tel: 071-987 6999
Fax: 071-538 4101

CIRIA Special Publication 97

Environmental Handbook for Building and Civil Engineering Projects

Volume 1: Design and specification

Checklists, obligations, good practice and sources of information

Principal authors

Roger Venables – Director, Venables Consultancy Services Ltd

David Housego – Associate Director, Phippen, Randall and Parkes

John Chapman – Associate, Phippen, Randall and Parkes

John Newton – Principal, Ecological and Environmental Services

Pamela Castle – Solicitor and Head of the Environmental Law Group, McKenna & Co

Ann Peirson-Hills – Solicitor, McKenna & Co

The authors have endeavoured to provide guidance to the legislation and the position on other matters as at 1 September 1993 with some minor later additions and amendments to 10 December 1993.

Summary

This Handbook contains information and practical guidance on the environmental issues likely to be encountered at each stage in the design and specification of a building or civil engineering project. It is aimed at informing anyone involved in such projects – for example, clients, engineers, design and other consultants, scientists and managers – on their obligations and the opportunities open to them to improve the industry's environmental performance.

Its layout has been designed to be accessible to a wide readership – from those seeking a concise overview or summary of relevant issues and legislation to others seeking more detailed guidance on accepted good practice. References to more-detailed guidance elsewhere are provided whenever possible.

The Handbook also forms a useful checklist and inventory of possible impacts and good practice for organisations considering setting up formal environmental management systems at a corporate or project level.

The main sections comprise:

- Agreeing the brief and setting the project environmental policy
- Inception and feasibility
- Primary design choices and scheme design
- Detailed design, working drawings and specifications
- Environmental considerations at tendering and contract letting.

Issues covered include:

- policy and forthcoming legislation
- energy conservation
- resources, waste minimisation and recycling
- pollution and hazardous substances
- internal environmental issues
- land use and conservation issues
- client and project team commitment.

A list of the main references is provided, together with many useful contact addresses and a checklist based on the contents list for individual readers to use on their projects.

This *Environmental Handbook* is one of a series of related documents and is published concurrently with an Environmental Handbook for the Construction Phase of Building and Civil Engineering Projects. Anyone involved in design and construction will need both volumes, though it is stressed that wherever an issue requires identical guidance to be given at the design and construction stages, the two volumes do present the same guidance in essentially the same way.

BREEAM (the Building Research Establishment Environmental Assessment Method) provides a procedure which enables a design or completed building to be assessed on the extent to which it addresses environmental issues. The different versions which currently cover new offices, supermarkets and superstores, existing offices and new homes will each be subject to periodic revision, reflecting progress in what constitutes good practice.

The *BSRIA Environmental Code of Practice for Buildings and Their Services* considers the impact of buildings from the viewpoint of the building services and provides a common strategic framework for the various disciplines involved. A draft document has been piloted and the final version is due for open publication in early 1994. The Code is a working document which makes recommendations on how to minimise the environmental impact of buildings over the entire building life cycle from conception through operation to eventual demolition.

Environmental Handbook for Building and Civil Engineering Projects
Volume 1: Design and Specification

Construction Industry Research and Information Association
Special Publication 97, 1994

Keywords:

Building, civil engineering, construction, environment, energy use, resources, materials, waste, recycling, pollution, handbook, good practice, hazardous substances, internal environment, design use, conservation, environmental legislation, environmental policy.

Reader interest:

Clients, developers, engineers, planners, design and other consultants, property owners and managers, Government, local authorities and regulatory bodies, academic and research organisations, legal and other advisers.

© CIRIA 1994
ISBN 0 86017 377 1
Thomas Telford ISBN 0 277 200 0

CLASSIFICATION	
Availability	Unrestricted
Content	Guidance based on good practice
Status	Committee Guided
User	Practitioners in property building and construction

Published by CIRIA, 6 Storey's Gate, Westminster, London SW1P 3AU, in conjunction with Thomas Telford Services Ltd, Thomas Telford House, 1 Heron Quay, London, E14 4JD.

Foreword

This Handbook has been prepared as part of the programme of the Construction Industry Environmental Forum and is aimed at providing practical guidance on environmental issues to clients, engineers, architects, builders, other designers, scientists and managers concerned with the design and/or specification of a building or civil engineering project. Its preparation is one of many initiatives within the construction industry to improve its environmental performance by identifying and responding to environmental issues in an appropriate manner. It was prompted by an earlier CIRIA project *Environmental issues in construction – A review of issues and initiatives relevant to the construction industry* which concluded that the lack of readily available guidance was a major impediment to improving the industry's environmental performance. This Handbook and a companion volume on the construction phase of building and civil engineering projects have been prepared to help the industry by providing background information and good practice guidance on a range of environmental issues.

The objectives in preparing the Handbooks have been:

- to provide a checklist for and guidance on the environmental considerations required at the various stages in the life of construction projects, identifying key activities and decision-making stages which can have significant impact on the environment;
- to produce a framework for the identification of existing information and to provide guidance, at an appropriate level of detail, of current good practice;
- to provide a framework to assist compilation by individual construction-related companies of registers of environmental effects and development of appropriate environmental management procedures, as required in BS7750: 1992 *Specification for Environmental Management Systems.*

The approach has been to provide succinct (rather than encyclopedic) guidance to good practice and obligations and to provide references to the source documents for detailed guidance where they are available. This Handbook addresses the main environmental issues that individual managers, engineers and scientists are likely to face in their day-to-day design and specification, whether they work for the client, designer, or specialist consultants, or for the main contractor or sub-contractors. The aim has been to be practical and to relate the guidance to the realities of present-day construction. Ease of use and access to the guidance have also been important considerations.

The extent to which environmental considerations can be built into any project will be influenced by a range of parties including the client, specifier, designer, developer and constructor and is sometimes constrained by cost and time. Whilst this Handbook is not intended as a formal code of practice, it indicates what is presently achievable given the necessary commitment, although it is accepted that not all the good practice guidance will be applicable on all projects because of client requirements, budget restraints and other factors.

The document also complements the activities of the Building Research Establishment and the Building Services Research and Information Association. The involvement of BRE and BSRIA in the production of this document has ensured that the guidance provided is broadly compatible.

This Handbook will help many in the construction and related industries to adopt good environmental practice. It must be stressed that there are many areas where practical guidance on actions that individuals can take to ensure good environmental performance is just not available. In such cases, we have sought to identify the nature of the issue and what limited approaches may be appropriate, and to indicate what further work is needed or under way to fill the gap in guidance. CIRIA would welcome feedback from readers on relevant issues so that future editions of the Handbooks and other reports can benefit from the industry's experience.

It must also be stressed that this Handbook has been prepared in the first two-thirds of 1993 against a rapidly developing industry scene. Whilst much of the guidance will hold good for a long time, readers must recognise the pace of change and take active steps to ensure they are using the latest editions of the quoted references and are up to date on the legislative position. It is recognised that in some important cases – the use of CFCs is a prime example – yesterday's good practice has become today's environmental problem. We all need to keep abreast of technical and economic developments to attain the objective of continuously improving our environmental performance.

The Handbooks have been prepared by a team led by Venables Consultancy:

- Roger Venables and Jean Venables at Venables Consultancy
- David Housego and John Chapman at Phippen, Randall and Parkes, Architects
- John Newton of Ecological and Environmental Services
- Pamela Castle and Ann Peirson-Hills of McKenna & Co, Solicitors, and
- Penny Mills, Book Designer,

with Paul Bartlett and Sandy Halliday representing the Building Research Establishment (BRE) and the Building Services Research and Information Association (BSRIA) as special advisors to the project. The team has endeavoured to summarise the accepted position on the issues covered as at September 1993 with some minor later amendments and additions.

The work has been guided by a Project Steering Group comprising representatives of CIRIA, BRE and BSRIA, contributors and specialists:

- Mr David Lush (Chairman) Ove Arup Partnership
- Mr David Adler Representing the Institution of Civil Engineers
- Mr Paul Bartlett Building Research Establishment
- Mr Peter Charnley National Westminster Bank Plc
- Mr Christopher Chiverell Laing Technology Group Ltd
- Mr Michael Gallon ICI Engineering
- Mr Ron German Stanhope Properties plc
- Ms Sandy Halliday Building Services Research and Information Association
- Mr Harry Hosker Building Design Partnership
- Mr Richard Jarvis, Environmental Division, W S Atkins Consultants Ltd
- Dr Alan Maries Trafalgar House Technology Ltd
- Mr John Troughton Construction Sponsorship Directorate, Department of the Environment.
- Dr Owen Jenkins CIRIA (Research Manager for the Project)

The work has been funded though the Construction Industry Environmental Forum with contributions from Department of the Environment, W S Atkins Consultants Ltd, Ove Arup & Partners, Building Design Partnership, ICI Engineering, the Institution of Civil Engineers, Laing Technology Ltd, National Westminster Bank Plc, Stanhope Properties Plc, Trafalgar House Technology, Lothian Regional Council and National House-Building Council.

CIRIA also wishes to express its gratitude to the following additional organisations who provided information and/or participated in a series of consultation meetings held during the course of the project: the Chartered Institute of Building, Chartered Institution of Building Services Engineers, Construction Industry Council, ECD Partnership, Association of Consulting Engineers, Building Employers' Confederation, Essex County Council, Great Mills Retail Ltd, Green Gauge Environmental Consultants, Galliford plc, David Lloyd Jones Associates, Miller Civil Engineers, National Rivers Authority, Oxford Brookes University, Geoffrey Reid Associates, University of Salford, Sheffield Hallam University, the Steel Construction Institute, Richards Moorehead & Laing, Scott Wilson Kirkpatrick & Partners, Sinclair Johnston Consulting Engineers and Taywood Environmental Consultancy.

How to use the Handbook

Introduction

This section on how to use the Handbook provides:

- an explanation of the structure of the Handbook;
- instructions on access to the guidance via the contents list (see page x) or the index (page 150);
- explanations of the terms used;
- explanations of the use of references to other documents.

Structure of the Handbook and access to the guidance

The Handbook is primarily structured by the stages in the design and specification process, from agreeing the brief to tender action and letting the construction contract(s), so that, knowing the stage in the process that you are presently engaged on, you can gain immediate access to the guidance on that stage by turning to the appropriate page given below:

Stage 1:	Agreeing the brief and setting the project environmental policy	3
Stage 2:	Inception and feasibility	11
Stage 3:	Primary design choices and scheme design	33
Stage 4:	Detailed design, working drawings and specifications	85
Stage 5:	Environmental considerations at tendering and contract letting	138

If, on the other hand, you wish to find out what the Handbook has to say on a particular environmental issue, irrespective of the stage in design and specification in which the issue may be of concern, turn to the Index on page 151, where you will find references to all sections where the issue is covered.

Each issue is dealt with in a box or table of two main types.

- The legislative position is reviewed in boxes with two main headings: ***Current legal position*** and ***Policy and forthcoming legislation***, each with references to match.

- The present technical position is reviewed in boxes with different headings: ***Background*** and ***Good Practice***, again with references to match. The ***Background*** section provides sufficient information and discussion of the issue for you to understand why the issue may be of concern and/or to explain why the issue is included in the Handbook. The ***Good Practice*** section provides summary guidance on what is accepted good practice or believed to be good practice, on the basis of existing reports, guidance documents and other publications that have been reviewed as part of the project leading to this Handbook.

Explanations of the issue box contents are provided in the diagrams on pages viii and ix. The Issue Titles and, in the technical boxes, the Good Practice Sections of each box are shaded.

References and further reading

The references quoted are of three main types:

- Background References – documents which you may choose to read to amplify the background information given but which are not essential for the full application of the handbook;
- Good Practice References – documents to which you ought to have ready access, either by ensuring that your office has a copy or by having a personal copy for ease of reference, because the Good Practice guidance makes specific reference to them for further detail;
- Legal References.

The main documents in the second and third groups are separately listed in the list of main references on page 140. Contact details of many information sources are given in the list of organisations on page 143.

How to use the Handbook – Legal issue boxes

The *Stage in the Design and Specification process* is given in the running header.
A Stage sub-heading is given above the first issue box in any new sub-stage.

The *Issue Title* is a topic of sufficient environmental concern that it merits discussion in a separate box.

The *Issue number* has up to four elements:
D identifies the issue as part of the Design and specification Handbook
First number is Stage in the Design and specification process.
Second number is the sub-stage or issue number.
Third number is the issue number when there are sub-headings.

The *Policy and forthcoming legislation* section sets out in broad terms a range of relevant developments such as government policy statements, developments in the EC and any of the issues that are expected to be subject to legislation in the relatively near future.

Current legal position is the main heading for information on and discussion of the legislation affecting the issues identified or the stage in design.
The *references to the current legal position* lists the main acts and regulations which you should at least be aware of and/or need to comply with when dealing with the issue concerned.

Policy references are any documents referred to in the policy section or which provide additional details of the issues raised.

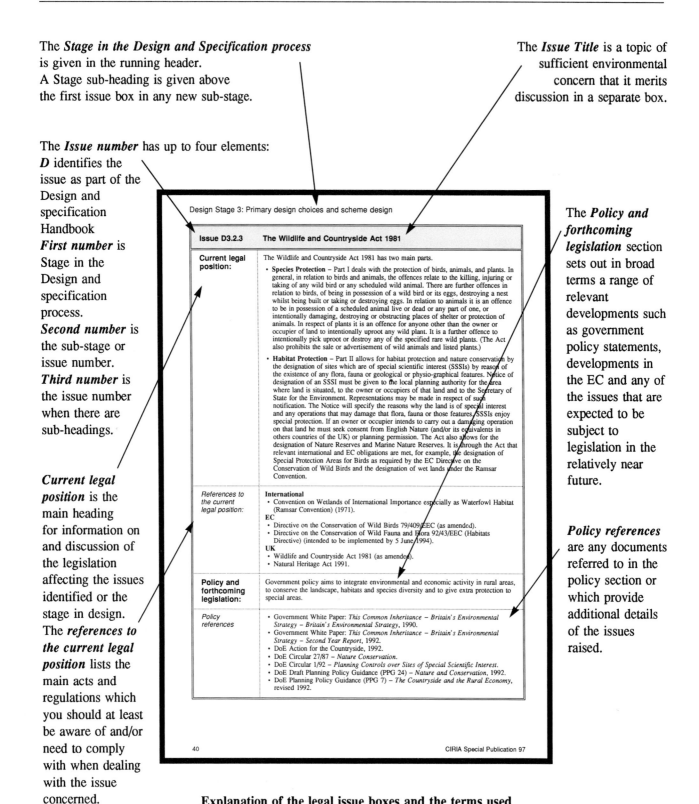

Design Stage 3: Primary design choices and scheme design

Issue D3.2.3	The Wildlife and Countryside Act 1981
Current legal position:	The Wildlife and Countryside Act 1981 has two main parts. • **Species Protection** – Part I deals with the protection of birds, animals, and plants. In general, in relation to birds and animals, the offences relate to the killing, injuring or taking of any wild bird or any scheduled wild animal. There are further offences in relation to birds, of being in possession of a wild bird or its eggs, destroying a nest whilst being built or taking or destroying eggs. In relation to animals it is an offence to be in possession of a scheduled animal live or dead or any part of one, or intentionally damaging, destroying or obstructing places of shelter or protection of animals. In respect of plants it is an offence for anyone other than the owner or occupier of land to intentionally uproot any wild plant. It is a further offence to intentionally pick uproot or destroy any of the specified rare wild plants. (The Act also prohibits the sale or advertisement of wild animals and listed plants.) • **Habitat Protection** – Part II allows for habitat protection and nature conservation by the designation of sites which are of special scientific interest (SSSIs) by reason of the existence of any flora, fauna or geological or physio-graphical features. Notice of designation of an SSSI must be given to the local planning authority for the area where land is situated, to the owner or occupiers of that land and to the Secretary of State for the Environment. Representations may be made in respect of such notification. The Notice will specify the reasons why the land is of special interest and any operations that may damage that flora, fauna or those features. SSSIs enjoy special protection. If an owner or occupier intends to carry out a damaging operation on that land he must seek consent from English Nature (and/or its equivalents in others countries of the UK) or planning permission. The Act also allows for the designation of Nature Reserves and Marine Nature Reserves. It is through the Act that relevant international and EC obligations are met, for example, the designation of Special Protection Areas for Birds as required by the EC Directive on the Conservation of Wild Birds and the designation of wet lands under the Ramsar Convention.
References to the current legal position:	**International** • Convention on Wetlands of International Importance especially as Waterfowl Habitat (Ramsar Convention) (1971). **EC** • Directive on the Conservation of Wild Birds 79/409/EEC (as amended). • Directive on the Conservation of Wild Fauna and Flora 92/43/EEC (Habitats Directive) (intended to be implemented by 5 June 1994). **UK** • Wildlife and Countryside Act 1981 (as amended). • Natural Heritage Act 1991.
Policy and forthcoming legislation:	Government policy aims to integrate environmental and economic activity in rural areas, to conserve the landscape, habitats and species diversity and to give extra protection to special areas.
Policy references	• Government White Paper: *This Common Inheritance – Britain's Environmental Strategy – Britain's Environmental Strategy*, 1990. • Government White Paper: *This Common Inheritance – Britain's Environmental Strategy – Second Year Report*, 1992. • DoE Action for the Countryside, 1992. • DoE Circular 27/87 – *Nature Conservation*. • DoE Circular 1/92 – *Planning Controls over Sites of Special Scientific Interest*. • DoE Draft Planning Policy Guidance (PPG 24) – *Nature and Conservation*, 1992. • DoE Planning Policy Guidance (PPG 7) – *The Countryside and the Rural Economy*, revised 1992.

40

CIRIA Special Publication 97

Explanation of the legal issue boxes and the terms used

How to use the Handbook – Technical issue boxes

The *Stage in the Design and specification process* is given in the running header.
A Stage sub-heading is given above the first issue box in any new sub-stage.

The *Issue number* has up to four elements with the same structure as with the legal issue boxes opposite.

Background is the main heading for information and discussion of the issue for you to understand why it is of concern and/or to explain why it is included in the handbook.

The *Issue Title* is a topic of sufficient environmental concern that it merits discussion in a separate box. It is repeated at the head of the second page of 2-page boxes.

Background references give details of any further reading that you should at least be aware of when dealing with the issue concerned.

The *symbol* ▸ is used to indicate other sections of this Handbook (D) or its companion volume on construction (C) which also contain relevant information or guidance.

The *Good Practice* section provides summary guidance on what the authors and the Project Steering Group have ascertained is accepted good practice or believe to be good practice. This has been developed on the basis of existing reports, guidance documents and other publications that have been reviewed as part of the project. The extent to which it will be possible to apply the guidance will vary from project to project.

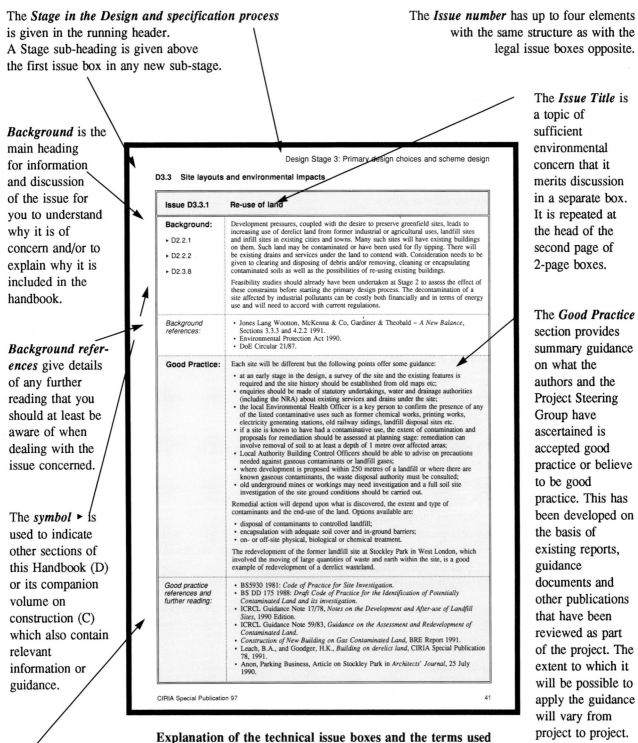

Design Stage 3: Primary design choices and scheme design

D3.3   Site layouts and environmental impacts

Issue D3.3.1	Re-use of land
Background: ▸ D2.2.1 ▸ D2.2.2 ▸ D2.3.8	Development pressures, coupled with the desire to preserve greenfield sites, leads to increasing use of derelict land from former industrial or agricultural uses, landfill sites and infill sites in existing cities and towns. Many such sites will have existing buildings on them. Such land may be contaminated or have been used for fly tipping. There will be existing drains and services under the land to contend with. Consideration needs to be given to clearing and disposing of debris and/or removing, cleaning or encapsulating contaminated soils as well as the possibilities of re-using existing buildings. Feasibility studies should already have been undertaken at Stage 2 to assess the effect of these constraints before starting the primary design process. The decontamination of a site affected by industrial pollutants can be costly both financially and in terms of energy use and will need to accord with current regulations.
Background references:	• Jones Lang Wootton, McKenna & Co, Gardiner & Theobald – *A New Balance*, Sections 3.3.3 and 4.2.2 1991. • Environmental Protection Act 1990. • DoE Circular 21/87.
Good Practice:	Each site will be different but the following points offer some guidance: • at an early stage in the design, a survey of the site and the existing features is required and the site history should be established from old maps etc; • enquiries should be made of statutory undertakings, water and drainage authorities (including the NRA) about existing services and drains under the site; • the local Environmental Health Officer is a key person to confirm the presence of any of the listed contaminative uses such as former chemical works, printing works, electricity generating stations, old railway sidings, landfill disposal sites etc. • if a site is known to have had a contaminative use, the extent of contamination and proposals for remediation should be assessed at planning stage: remediation can involve removal of soil to at least a depth of 1 metre over affected areas; • Local Authority Building Control Officers should be able to advise on precautions needed against gaseous contaminants or landfill gases; • where development is proposed within 250 metres of a landfill or where there are known gaseous contaminants, the waste disposal authority must be consulted; • old underground mines or workings may need investigation and a full soil site investigation of the site ground conditions should be carried out. Remedial action will depend upon what is discovered, the extent and type of contaminants and the end-use of the land. Options available are: • disposal of contaminants to controlled landfill; • encapsulation with adequate soil cover and in-ground barriers; • on- or off-site physical, biological or chemical treatment. The redevelopment of the former landfill site at Stockley Park in West London, which involved the moving of large quantities of waste and earth within the site, is a good example of redevelopment of a derelict wasteland.
Good practice references and further reading:	• BS5930 1981: *Code of Practice for Site Investigation*. • BS DD 175 1988: *Draft Code of Practice for the Identification of Potentially Contaminated Land and its investigation*. • ICRCL Guidance Note 17/78, *Notes on the Development and After-use of Landfill Sites*, 1990 Edition. • ICRCL Guidance Note 59/83, *Guidance on the Assessment and Redevelopment of Contaminated Land*. • *Construction of New Building on Gas Contaminated Land*, BRE Report 1991. • Leach, B.A., and Goodger, H.K., *Building on derelict land*, CIRIA Special Publication 78, 1991. • Anon, Parking Business, Article on Stockley Park in *Architects' Journal*, 25 July 1990.

CIRIA Special Publication 97   41

Explanation of the technical issue boxes and the terms used

Good practice references and further reading may give details of documents needed to secure the detail on the guidance given and/or be suggested further reading if you wish to delve more deeply into current practice on the issue concerned.

Contents

Summary   iii

Foreword   iv

How to use the Handbook   vii

List of abbreviations used   xiii

Introduction   1

Stage 1 Agreeing the design brief and setting the project environmental policy

    1.1 Gaining commitment   4

    1.2 Agreeing the design brief   6

    1.3 Setting environmental policy
        1.3.1 Consultants' potential influence on the environmental aspects of a project   7
        1.3.2 Setting environmental policy and the possible role of environmental management
                systems   8

    1.4 Briefing the design team and assigning responsibility   10

Stage 2 Inception and feasibility

    2.1 Legislation and policy   12

    2.2 Environmental assessment   14

    2.3 Site issues
        2.3.1 Derelict and contaminated land   16
        2.3.2 Site hydrology and water quality   18
        2.3.3 Flora, fauna and landscape   20
        2.3.4 Landfill gas and naturally-occurring methane and CO_2 22
        2.3.5 Non-ionizing radiation   23
        2.3.6 Radon and thoron   24
        2.3.7 Noise   25
        2.3.8 Services infrastructure   26
        2.3.9 Local community   26
        2.3.10 Local climate – assess impact of the local climate on development   28
        2.3.11 Local transport infrastructure   29
        2.3.12 Cultural features   30

    2.4 Securing outline planning permission and identifying financial implications   31

Stage 3 Primary design choices and scheme design

    3.1 Briefing the design team   35

    3.2 Legislation and policy
        3.2.1 Planning legislation   36
        3.2.2 Building Regulations   38
        3.2.3 The Wildlife and Countryside Act 1981   40

    3.3 Site layouts and environmental impacts
        3.3.1 Re-use of land   41
        3.3.2 Ecological value of site   42
        3.3.3 Transport and access to site   43
        3.3.4 Local transport infrastructure   44
        3.3.5 Building microclimate   45
        3.3.6 Overshadowing and access to daylight and sunlight   46

3.3.7	Passive solar design and its effect on site layouts	47
3.3.8	Earth sheltered design	48

✓ 3.4 Impacts on occupants and users, visitors and neighbours
3.4.1	User and community consultation	49
3.4.2	External appearance (aesthetics)	50
3.4.3	Noise	51
3.4.4	Anticipating and minimising development impacts	52

✓ 3.5 Detailed consultation with relevant bodies
3.5.1	Water quality and economy	52
3.5.2	Floodwater provisions	53
3.5.3	Archaeological and historical issues	54

3.6 Designing appropriate landscaping   54

✓ 3.7 Energy options
3.7.1	Energy conservation and efficiency	56
3.7.2	Use of high levels of insulation to minimise energy use in buildings	58
3.7.3	Passive solar design of domestic buildings	59
3.7.4	Renewable energy sources	60
3.7.5	Combined heat and power	60
3.7.6	Design of civil engineering works for minimal energy use	61

✓ 3.8 Labelling and other environmental information
3.8.1	Use of energy labelling schemes	62
3.8.2	Green labelling of materials and products	63
3.8.3	BREEAM	64

✓ 3.9 Ventilation options in buildings
3.9.1	Air infiltration and ventilation rates	66
3.9.2	Natural ventilation and passive stack ventilation in small buildings	67
3.9.3	Ventilation in large structures and buildings	68
3.9.4	Mechanical ventilation and heat recovery	69
3.9.5	Air conditioning	70
3.9.6	Legionnaires' Disease	71

3.10 Daylight and need for artificial light   72

✓ 3.11 Occupant comfort   73

3.12 Criteria for primary material selection
3.12.1	Material for the primary structure	74
3.12.2	Brick, block and masonry construction	75
✓ 3.12.3	Cladding materials	76
3.12.4	Roofing materials	77
3.12.5	Timber	78
✓ 3.12.6	Services and materials	79
3.12.7	Insulants	80

✓ 3.13 Use of materials
3.13.1	Waste, salvage, re-use and recycling of materials	81
3.13.2	Design to minimise use of materials	82
3.13.3	Storage of recyclable materials	83
3.13.4	Design for maintenance and cleaning	84

Stage 4 Detailed design, working drawings and specifications

    4.1   Ensuring the design team knows the project environmental policy   86

    4.2   Legislation and policy
        4.2.1   Building Regulations (As 3.2.2 but repeated for ease of use)   86
        4.2.2   Avoidance of hazardous materials   88
        4.2.3   Radon and thoron – legal issues   90
        4.2.4   Radon, thoron and naturally-occurring methane – technical issues   91
        4.2.5   Contaminated land and toxic substances   92
        4.2.6   Electromagnetic radiation – legal issues   94
        4.2.7   Electromagnetic radiation – technical issues   95
        4.2.8   Safety, security and fire   96

    4.3   Use of energy
        4.3.1   Heating   98
        4.3.2   Cooling   99
        4.3.3   Insulation   101

    4.4   Labelling and other environmental information
        4.4.1   Energy labelling schemes   102
        4.4.2   BREEAM   103

    4.5   Landscape, ecology and the use of plants
        4.5.1   Landscape and ecology   104
        4.5.2   Use of plants to 'green' the built environment   105

    4.6   Internal environment
        4.6.1   Services   106
        4.6.2   Water use and disposal   107
        4.6.3   Noise and acoustics   108
        4.6.4   Ventilation and air quality   110
        4.6.5   Lighting   112
        4.6.6   Condensation   114
        4.6.7   Controls   115
        4.6.8   Legionnaires' Disease   116
        4.6.9   Sick Building Syndrome   118

    4.7   Materials
        4.7.1   Use of environmentally acceptable materials   120
        4.7.2   Environmental policy on materials   121
        4.7.3   Criteria for selection   122
        4.7.4   Avoidance of CFCs, HCFCs and reduction in NO_x, CO_2 and SO_x 124
        4.7.5   Sourcing of timber   126
        4.7.6   Transport of materials and components   127
        4.7.7   Vetting suppliers   128
        4.7.8   Use of recycled materials   130

    4.8   End-use considerations
        4.8.1   Environmental issues in respect of end-use   132
        4.8.2   Operational and post-operational requirements   133
        4.8.3   Commissioning   134

    4.9   Handover of project environmental policy to contractor and/or other consultants   135

Stage 5 Environmental considerations at tendering and contract letting   138

Main background, good practice and legislation references   140

Addresses of relevant organisations   143

Environmental Checklist – Design   147

Index   151

List of abbreviations used

AECB	Association for Environment Conscious Building
AONB	Area of Outstanding Natural Beauty
BEC	Building Employers' Confederation
BEMS	Building energy management systems
BEPAC	Building Environmental Performance Analysis Club
BRE	Building Research Establishment
BRECSU	Building Research Energy Conservation Support Unit
BREEAM	Building Research Establishment Environmental Assessment Method
BS7750	British Standards Institution *Specification for Environmental Management Systems*
BSI	British Standards Institution
BSRIA	Building Services Research and Information Association
CBI	Confederation of British Industry
CHP	Combined Heat and Power
CHPA	Combined Heat and Power Association
CIBSE	Chartered Institution of Building Services Engineers
CIC	Construction Industry Council
CIEC	Construction Industry Employers' Council
CIEF	Construction Industry Environmental Forum
CIOB	Chartered Institute of Building
CIRIA	Construction Industry Research and Information Association
CFCs	Chlorofluorocarbons
COPA	Control of Pollution Act 1974
COSHH	The Control of Substances Hazardous to Health Regulations, 1988
CPRE	Council for the Protection of Rural England
DoE	Department of the Environment
DoT	Department of Transport
DTI	Department of Trade and Industry
EA	Environmental Assessment
EC	European Community, Environment Council
EDAS	Energy Design Advisory Scheme
EEO	Energy Efficiency Office
EMS	Environmental Management System
EPA	Environmental Protection Act 1990
ESTA	Energy Systems Trade Association
ETSU	Energy Technology Support Unit

FCEC	Federation of Civil Engineering Contractors
HAPM	Housing Association Property Mutual Ltd
HCFCs	Hydrochlorofluorocarbons
HFCs	Hydrofluorocarbons
HMSO	Her Majesty's Stationery Office
HSC	Health and Safety Commission
HSE	Health and Safety Executive
HSWA	The Health and Safety at Work etc Act 1974
HVCA	Heating and Ventilation Contractors Association
ICE	The Institution of Civil Engineers
ICRCL	Interdepartmental Committee for the Redevelopment of Contaminated Land
IEE	Institution of Electrical Engineers
IMECHE	Institution of Mechanical Engineers
IWEM	Institution of Water and Environmental Management
NEDO	National Economic Development Office
NEF	National Energy Foundation
NHBC	National House-Building Council
NHER	National Home Energy Rating Scheme
NRA	National Rivers Authority
NRPB	National Radiological Protection Board
OJ	Official Journal of the European Commission
RIBA	Royal Institute of British Architects
RICS	Royal Institution of Chartered Surveyors
RSNC	Royal Society for Nature Conservation
RSPB	Royal Society for the Protection of Birds
SBS	Sick Building Syndrome
SI	Statutory Instrument or Site Investigation
SNCI	Site of Nature Conservation Importance
SP	(CIRIA) Special Publication
SSSI	Site of Special Scientific Interest
TPO	Tree Preservation Order
TRADA	Timber Research and Development Association
UK	United Kingdom (of Great Britain and Northern Ireland)
WWF	World Wide Fund for Nature

Introduction

Environmental issues in construction

The UK Government set out its environmental aims in the White Paper *This Common Inheritance* and, following the 1992 Rio de Janeiro conference on the environment and development, is committed to the production of a Sustainable Development Plan for which a Consultation Paper was issued in July 1993. As well as these Government imperatives, there are also pressures for environmental improvement at European Community, national and local levels, from pressure groups, clients and employees as well as the general public. Wide-ranging changes in attitudes towards the environment are taking place, prompted partly by legislation, partly by client influence in the market place, and partly through personal decisions made by individuals at work and in their day-to-day lives.

The industry's response has been extensive, both in reviewing the key issues it faces and in developing research programmes aimed at producing guidance. The response has included:

- the establishment of the Construction Industry Environmental Forum by CIRIA, BRE and BSRIA;
- research studies by these and other research bodies;
- initiatives led by the Professional Institutions and Trade Associations;
- the establishment of the Institute of Environmental Assessment;
- corporate initiatives.

These initiatives have included:

- participation in the trial use of BS7750: *Specification for Environmental Management Systems* by construction industry sector groups;
- the development of corporate environmental policies with targets to improve environmental performance;
- BRE's Environmental Assessment Method for different types of buildings (BREEAM);
- BSRIA's *Environmental Code of Practice for Buildings and Their Services*, which has been piloted and is due for open publication in early 1994;
- the CIC Environment Task Group's report *Our land for our children: an environmental policy for the construction professions*, published in August 1992;
- the CIEC Environment Task Force's report, *Construction and the Environment*, published by the Building Employers' Confederation in May 1992.

In addition, other groups have launched more general, business-related initiatives which seek to secure commitment to continuous improvement in environmental performance by participating companies.

Very many environmental improvements cost nothing to implement except thought, knowledge and commitment, whilst many can also provide cost savings and increase the attractiveness of adopting a positive environmental stance. However, it must also be recognised that some highly desirable environmental improvements will incur additional costs to one or more of the parties involved in a project. The commercial position of the construction industry is rarely if ever conducive to the imposition of such additional costs, whether to clients, designers, contractors or suppliers. As a result, an extra dimension has been added to the environmental challenge – that of improving environmental performance whilst generating new market opportunities and enhancing competitiveness.

The role of designers, contractors and clients

The design professions can and should have a very significant influence on the environmental impact of the projects they design and develop. In broad terms, the research undertaken in preparing these Handbooks indicates that there is adequate information available for a designer who wishes to pursue a policy of minimising the environmental impact of a design: to do so will require commitment and extra effort, but it is not only possible but necessary.

Those responsible for the construction phase must also seek to reduce the impact of their own operations on the environment. Their role is to translate a design into an operational scheme so their ability to redress any inherent environmental weaknesses in a design is limited. In contrast to designers, those responsible for the construction phase can find out *what* they *ought* to do as environmentally responsible contractors but guidance on *how* to achieve it is comparatively lacking. Reference to the guidance in this Handbook will assist in this endeavour.

Clients of the construction industry can also have a marked influence on the environmental performance of their projects. Some clients will not be looking for enhanced environmental performance: the design professions have a particularly important role in persuading such clients towards a more-environmentally responsible approach.

Key issues

From the very wide range of environmental issues facing the industry, a few stand out as very important:

- energy conservation and the need to reduce emissions of so-called greenhouse gases;
- selection of materials against environmental criteria;
- effective conservation of natural resources – designing and constructing with waste minimisation, recycling and salvage in mind;
- pollution control;
- bringing derelict or contaminated sites into beneficial use;
- securing commitment from all concerned to a more environmentally responsible approach;
- the move towards a sustainable environment, an approach which embodies a positive, constructive adoption of the above issues.

Coverage of the Handbook

Readers should note four main points about the coverage of the guidance in this Handbook.

- Many environmental issues affect or are affected by related occupational health and safety concerns. Whilst the impact of the 1974 Health & Safety at Work Act is introduced and its main implications stressed, the Handbook only covers occupational health and safety issues where they have a significant wider environmental impact or may affect the working environment in works or buildings. Mainstream occupational health and safety of construction workers is not covered.

- On the current legal position, the UK situation mainly refers to England and Wales. Generally, legislation which applies to England and Wales also applies to Scotland, with such legislation being modified to take account of the different administrative and judicial systems in Scotland. Where relevant, Scottish primary legislation is listed if it is different from that applying to England and Wales. However, detailed differences in Scottish law are not discussed. No reference is made to Northern Ireland which for the most part has different environmental laws and administrative systems from England and Wales. However, Box 8.9 in *Environmental issues in construction – A review of issues and initiatives relevant to the construction industry* (CIRIA Special Publications 93 and 94 1993) gives a listing of the main environmental legislation in England and Wales, Scotland and Northern Ireland.

- A distinction must be drawn between legislative and regulatory requirements and achievable good practice which can, sometimes, go beyond legal minimum standards of environmental performance. The requirements of the former are explicit in the coverage given to relevant regulations. The accepted good practice which is described may also be derived from regulatory requirements. However, in many cases it will be drawn from industry codes of practice, and/or formed from general consensus within the industry and relevant groups, which indicate that higher standards of performance are achievable. Best practice is not defined, since there may be many project-specific criteria which dictate the suitability of the many possible solutions to a problem. It must also be acknowledged that as the regulatory framework is subject to change, so is good practice.

- The Handbook is equally directed at building and civil engineering. Whilst readers will notice that some guidance is specifically related to building design and construction and some to civil engineering schemes, the principles applied are often interchangeable.

Stage D1   Agreeing the design brief and setting the project environmental policy

Stage D1: Agreeing the design brief and setting the project environmental policy reviews the reasons for being committed to an environmental performance better than that demanded by regulatory and legislative measures, and the potential benefits arising from such a commitment. Turning it into environmental objectives and policy, and possibly into an environmental management system, is then addressed together with the need to designate a member of staff with the responsibility for policy implementation, and to ensure that all members of the project team understand its implications and are committed to its realisation.

D1.1   Gaining commitment

D1.2   Agreeing the design brief

D1.3   Setting environmental policy

D1.4   Briefing the design team and assigning responsibility

D1.1 Gaining commitment

Issue D1.1	Gaining commitment
Background:	*Introduction* – The construction industry has a major impact on the natural and built environments at all levels: local, national and global. This presents it with the opportunity to take the lead in demonstrating industry's contribution to solving environmental problems, with some of the solutions being directly or indirectly applicable to other industries. To take this lead in an effective and rigorous way, individuals involved need to have a basic understanding of the issues involved and a commitment to environmental improvement.

Whilst legal and regulatory instruments will set minimum standards which will continue to be revised, added to and improved upon, they will almost inevitably lag behind technological advances, accepted knowledge or areas of increasing concern or action. For example, techniques for radically improving upon energy efficiency levels set by building regulations are well known and readily attainable. A commitment to doing better will result in many environmental advantages and will demonstrate to society the willingness of the construction industry to improve its environmental performance. It must be acknowledged that adopting an environmentally responsible approach may create commercial advantages or disadvantages depending on the particular markets, but there are many benefits from improving environmental performance.

- *Legislation and regulation* – Being fully aware of the legislative position will ensure compliance with environmental legislation and regulations, thus avoiding fines and clean-up costs.
- *Market forces* – keeping one step ahead will ensure that an individual company is able to respond more quickly and effectively to any changes which threaten to influence the market place. Increased concern for the environment has also meant that a sizable proportion of the market is now looking for companies, products and services which in some way demonstrate a commitment to minimising environmental damage. Purchasers and occupiers of buildings are starting to look at energy efficiency as a means of keeping fuel bills under control in a 'carbon conscious' future. Environmental rating schemes for buildings may make those with a 'green certificate' more attractive to customers and may in the long term improve rental performance.
- *Avoiding problems* – environmental issues are generating wider concern. The implications for industry can include delays in completing projects, unforeseen need for specialist advice, plant which fails to comply with new standards, poor community relations or a product nobody wants. A commitment to and awareness of environmental issues should ensure that these problems are avoided or minimised.
- *Staff recruitment and morale* – individuals are looking for ways of improving environmental quality not only at home but also at work. Many want to see an improvement in their company's environmental performance and may even refuse to work for a company with a poor environmental record. A poor environmental record and accompanying bad publicity can affect staff morale and act as a barrier to quality recruitment.
- *Financial* – project planning and management with the environment in mind can reduce waste and ensure more appropriate design. However, new skills will be needed and a different approach to that normally employed may be necessary. When in a competitive situation regarding fees it may be that extra costs have to be incurred for acquiring these skills and adopting new approaches.
- *Self interest and altruism* – the human race is part of the environment, and environmental improvement will bring benefits to everyone. Individuals stand to gain from their own efforts to minimise their or their company's impact on the environment and other people and the environment will then benefit as well. There is, however, always likely to be some conflict between market forces and an altruistic approach to improving environmental performance. |
| *Background references:* | - CIC Environment Task Group, *Our land for our children: An environmental policy for the construction professions*, Construction Industry Council, August 1992.
- CIEC Environment Task Force, *Construction and the Environment*, Building Employers' Confederation, May 1992.
- *Environmental issues in construction – A review of issues and initiatives relevant to the building, construction and related industries*, CIRIA Special Publications 93 and 94, 1993.
- Jones Lang Wootton, McKenna & Co., and Gardiner & Theobald – *A New Balance: Buildings and the Environment – A Guide for Property Owners and Developers*, Chapters 1 and 2, 1991. |

Issue D1.1	Gaining commitment

Good Practice: ▶ D1.3.2	*In reaching an agreed commitment to a positive project environmental policy*, clients, their consultants and, if appropriate, their design-and-build contractors should consider and recognise the importance of awareness, setting an environmental policy, assigning responsibility and the possible adoption of a formal environmental management system (EMS). *In particular, they should consider or recognise that:* • taking a lead in this area requires construction-related organisations to have a commitment to environmental improvement made at board level (or its equivalent) with the development and publication of a corporate environmental policy statement; • a basic awareness and understanding of environmental issues is necessary before any commitment to environmental improvement can be made; • joining an organisation such as the CIRIA/BRE/BSRIA Construction Industry Environmental Forum can be extremely beneficial in enabling discussion of the various issues with fellow practitioners, in providing opportunities to gain from the knowledge of specialists and experts and to be made aware of new issues to consider; • attendance at workshops or conferences can be useful, although a more specific training course aimed at raising environmental awareness may be more effective; • there is a wealth of written material available on environmental matters to refer to, the most well-known being included in the reference lists in this document; • British Standards Institution, BS7750: 1992 *Specification for Environmental Management Systems* provides a framework for the establishment of an environmental management system which could be applied to a single, environmentally sensitive project; • whether a full environmental management system is to be adopted or not, there are benefits in assigning responsibilities for environmental matters to one individual and assigning the title Project Environmental Manager (though the post does not of course have to a full-time job); • having acquired the necessary information, the depth and extent of the commitment to environmental improvement can then be decided upon; • setting down the agreed commitment in a short environmental project policy statement which could be made available to staff and other personnel involved in a particular project will be extremely beneficial in securing the necessary commitment from all involved; • such a commitment may go as far as the client chooses, although it should not be so ambitious as to be unrealistic whilst, on the other hand, there would be little point in just accepting the legislative and regulatory requirements; • consideration needs to be given to how environmental commitment is to be demonstrated, for example, by appointing staff or carrying out research in an effort to improve performance, or by employing better practice in project implementation, or by sponsoring environmental initiatives off site or making staff or resources available to environmental groups; • environmental commitment, policy statements and procedures should be periodically reviewed by senior managers to judge their effectiveness and such environmental policies and practices should be upgraded based on the review's results; • without explicit environmental commitment, any attempts to tackle environmental problems on the project are likely to be ill-considered and ultimately ineffective.
Good practice references and further reading:	• As background references, plus: • Miller, S., *Going Green*, JT Design Build, Bristol, undated. • Leicester County Council, *Building for the Environment: An Environmental Good Practice Checklist for the Construction and Development Industries*, Leicester County Council jointly with Leicester City Council, November 1992. • Halliday, S.P., *Environmental Code of Practice for Buildings and Their Services*, BSRIA, 1994. • BS7750: 1992 *Specification for Environmental Management Systems*, BSI, Milton Keynes. • Jones Lang Wootton, McKenna & Co., Gardiner & Theobald, *A new balance*, Chapters 2 and 3, 1991. • Prior, J.J., and Raw, G.J., *The Environment and the Construction Industry*, Building Research Establishment, 1992.
Legal references:	• Government White Papers: *This Common Inheritance – Britain's Environmental Strategy*, and *Yearly Reports*, 1991 and 1992.

D1.2   Agreeing the design brief

Issue D1.2	Agreeing the design brief
Background:	In any construction project many people will have an interest in and potential influence over its execution and outcome. These 'stakeholders' could include clients, designers, local community, pressure groups etc. Ensuring that the scope and nature of the assignment is agreed in some detail by the parties involved is a major step towards commitment and will help avoid misunderstandings in the future.
Background references:	• *Environmental Assessment: Guide to the identification, evaluation and mitigation of environmental issues in construction schemes*, CIRIA Special Publication 96, 1993. • Leicester County Council, *Building for the Environment: An Environmental Good Practice Checklist for the Construction and Development Industries*, Leicester County Council jointly with Leicester City Council, November 1992. • Miller, S., *Going Green*, JT Design Build, Bristol, undated. • Therivel, R. et al, *Strategic environmental assessment*, Earthscan Publications, London, 1992.

Good Practice:	*A statement of the environmental aspects of a design brief should*:
	
	• be presented in documents and at meetings in a way that can be understood by all those to be involved in it and who may be affected by it;
	• in most cases, be in the form of a comprehensive, easy-to-read and fully illustrated statement, whose main points are presented at meetings of selected interested parties;
	• include descriptions of:
	  – the proposed type of construction;
	  – the perceived end use and contribution to the local area;
	  – the need for a preliminary environmental assessment to identify impacts and possible mitigations;
	  – the need for a formal Environmental Assessment and publication of an environmental statement;
	  – any more general statements of the environmental aims or objectives for the project, for example, whether or not it is intended to submit the completed project for any environmental rating scheme (e.g. BREEAM or NHER/Starpoint energy rating) or if an corporate environmental management system, such as one based on BS7750, will be applied to the project;
	• be as brief as possible whilst providing a reasonably full treatment of the issues involved.

Good practice references and further reading:	• *Design Briefing Manual*, AG1/90, BSRIA, 1990. • Prior, J.J., (Editor), BREEAM 1/93, *An environmental assessment for new offices*, BRE, 1993. • Crisp, V.H.C, Doggart, J., and Attenborough, M., BREEAM 2/91, *An environmental assessment for new superstores and supermarkets*, BRE, 1991. • Prior, J.J., Raw, G.J. and Charlesworth, J.L., BREEAM 3/91, *An environmental assessment for new homes*, BRE, 1991. • Baldwin, R., Bartlett, P., Leach, S.J. and Attenborough, M., BREEAM 4/93, *An environmental assessment for existing office buildings*, BRE, 1993.
Legal references:	• See D2.1 and D2.2.

D1.3 Setting environmental policy

Issue D1.3.1	Consultants' potential influence on the environmental aspects of a project
Background: ▸ D1.1 ▸ D1.2	In some situations the driving force for improving the environmental performance of a project may not come from the client but from one of the consultants on their team. In these instances the consultant has the task of influencing the client in respect of environmental issues using the types of arguments mentioned in D1.1 and by means based on D1.2. Active marketing of the advantages accruing from improved environmental performance is needed if consultants are to seek a market advantage over competitors for their environmental expertise.
Background references:	• Halliday, S.P., *Environmental Code of Practice for Buildings and Their Services*, BSRIA, 1994. • Construction Industry Environmental Forum, *Environmental issues in construction – A review of issues and initiatives relevant to the building, construction and related industries*, CIRIA Special Publications 93 and 94, 1993.
Good Practice:	*Wherever possible, consultants should stress:* • the advantages of addressing environmental issues in a thorough and rigorous way; • that they are able to provide the necessary skills, expertise and advice; • that environmental quality is something worth having and that this can be recognised in the fees which are agreed.
Good practice references and further reading:	No specific references identified.

Issue D1.3.2	Setting environmental policy and the possible role of environmental management systems

Background:	Publication of an environmental policy statement is increasingly being seen as demonstrating the extent of commitment to environmental issues and the starting point for improvement in performance.An environmental policy comprises a series of short statements which state the aims and principles of action of a company or project in relation to the environment, given the constraints it is working under at any specific time.
	An environmental management system (EMS) defines, regulates and monitors how a company puts its corporate environmental policy into practice either on a company-wide basis or a project-by-project basis. It enables the company to manage its own environmental performance in a logical step-by-step approach.
	BS7750, the British Standards Institution Specification for Environmental Management Systems, published in 1992, sets out how a company-wide EMS may be set up and follows very similar principles to BS5750 on Quality Systems. In addition, the EC Eco-management and audit regulation sets out the basis for an environmental management and audit system.
	An environmental policy statement should ideally be a public document and must be made public if any EMS set up to implement the policy is to meet the requirements of BS7750. It is necessary, before drawing up an environmental policy, to reflect on the scope of a company's or project's likely influence on the environment. BS7750 requires the preparation of a register of environmental effects which will ensure that all areas of the company's operations will be covered by the policy statement and will help to ensure that it is realistic in its objectives.
	BS7750 has been piloted by a number of industry sector groups, including two related to construction. A draft second edition was then issued in mid-1993 for open comment, the deadline being 31 August 1993. A new edition is to be produced but no date has yet been set. It is therefore too early to expect mainstream consultants and contractors to have BS7750-based EMSs in place but it may be appropriate to consider setting up a project-based EMS based on the principles of BS7750.
	In the proposed development and implementation of BS7750, it is planned to establish certification schemes similar to those under BS5750 for Quality Systems. It is therefore envisaged that independent accreditors will be appointed to verify that the management system has been developed in accordance with the requirements of BS7750, and then to audit that it has been rigorously followed and environmental improvements have been made. The EMSs envisaged will not give a company or project a 'mark out of ten' for environmental performance but will be aimed at securing continued environmental improvements. EMSs developed in accordance with BS7750 will therefore have the advantage that a company will be able to receive BS7750 accreditation knowing full well that it has some way to go to achieving its best possible environmental performance; at least it will have the confidence that it is tackling the problems in an effective and approved way.
	As BS7750 is still in its formative stages, putting an EMS that meets its requirements into practice and gaining accreditation may prove to be a lengthy process. However, the underlying principles are sound and could be adopted to suit a particular organisation's own aims and objectives. As with quality systems developed using BS5750 as a basis but which are not independently assessed, EMSs could be developed by individual companies to suit their requirements by using BS7750 as the basis but without the intention of seeking independent assessment.
Background references:	**UK** • BS7750: 1992 *Specification for Environmental Management Systems*, BSI, Milton Keynes. • BSI, Draft second edition of BS7750, Document DC400220/93, BSI, London, 1993. • Laing Technology, *The Laing Environment – Environmental Policy Statement and practice notes*, 1990 onwards. • *Corporate environmental policy statements*, Confederation of British Industry, 1992. • Croners, *Environmental Management*, with quarterly amendment service, Croner Publications Ltd, First Edn, October 1991. • *Environment Business Forum Agenda for Voluntary Action*, CBI, undated. **EC** • Regulation for voluntary participation by companies in the industrial sector in a community eco-management and audit scheme (Agreed 23 March 1993).

Issue D1.3.2	Setting environmental policy and the possible role of environmental management systems

| Good Practice:

▶ D4.9 | What comprises good practice in the setting of environmental policy is changing rapidly as more construction organisations develop and publish corporate environmental policy statements as required by BS7750 and as experience is gained on how to deal with many of the environmental pressures facing the construction industry.

The key features of a corporate or project environmental policy include:

• commitment of the board (or its equivalent) and of senior managers;
• regular measurements of performance and reviews of progress in developing policy and procedures by an executive committee;
• coverage of the following issues:
   – energy;
   – resources and materials (including replacements for CFCs);
   – transport issues;
   – pollution (air/water/soil);
   – noise;
   – landscape and ecology;
   – cultural issues;
   – good neighbour issues;
   – whether or not to follow an EMS or to ask for a BREEAM assessment;
   – environmental impact of suppliers;
   – waste management;
   – expectations of staff;
   – corporate/public relations and the environment;
• compliance with relevant regulations;
• clear assignment of responsibilities for implementation, monitoring, audit, review and revision of the system.

The stages in the development and implementation of an environmental management system are set out in BS7750 as:

• secure commitment;
• initial review;
• set policy;
• set organisation and appoint personnel;
• establish register of regulations;
• evaluate environmental effects and establish register;
• set objectives and targets;
• set up management programme and write management manual;
• set up and implement operations control;
• keep records;
• undertake audits and reviews;
• review and re-set policy.

It is hoped that the structure and content of this handbook, and the companion on the construction phase, will assist managers of construction-related organisations in the preparation of the register of environmental effects and in developing their environmental management system. |
| *Good practice references and further reading:* | • As background references, plus:
• BS7750: 1992 *Specification for Environmental Management Systems*, British Standards Institution, Milton Keynes.
• Halliday, S.P., *Environmental Code of Practice for Buildings and Their Services*, BSRIA, 1994.
• Construction Industry Environmental Forum, *Environmental management in the construction industry*, Notes of meeting held on 22/09/92, CIRIA, 1992.
• CIC Environment Task Group, *Our land for our children: an environmental policy for the construction professions*, Construction Industry Council, August 1992.
• Miller, S., *Going Green*, JT Design Build, Bristol, undated.
• Dauncey, G., Bamberton – a town for the future, in *Ecodesign*, Vol II No.1. pp32–33 – a background article on the development of an environmental code and design principles for this proposed new town in Canada. |

D1.4 Briefing the design team and assigning responsibility

Issue D1.4	Briefing the design team and assigning responsibility
Background: ▸ D5	The construction industry incorporates a great range of skills and expertise, and depends on the acquisition and dissemination of a wide variety of information from the policy, legal or regulatory to that of a scientific or technical nature. The environmental issues encountered by the construction industry are similarly complex and inter-related. Effective communication between the different parties can therefore save time and money and ensure the successful realisation of a project's aims and objectives. Design team members will need to consider a wide range of issues and bear in mind the implications of their work at all stages, from the construction phase to operational use of the scheme through to demolition. To ensure that the design team work in an effective and integrated way and are aware of the consequences of their environmental decisions, it is critical that the whole team is adequately briefed and are committed to working together in a team. It needs to be aware of all the issues involved, the degree of commitment made to solving them, and the policies and programmes to be implemented. Information will need to be shared amongst team members and passed to contractors and others involved in the construction process in order that appropriate decisions and actions are carried out in full knowledge of their consequences. Ill-informed decisions made at this stage can result in penalties in the future. Every member of that team should be committed to the project environmental policy, although it is wise to assign to one person – perhaps the design team leader or an environmental specialist – the responsibility for ensuring that policies and procedures are carried through. Such a person, who may be assigned the title of Environmental Manager, could then take on the responsibility for providing relevant documentation to the team and act as a repository of information and guidance on environmental matters. Each discrete team within the whole construction process should have one person who is ultimately responsible for environmental issues even though they may not necessarily be a technical expert in any environmental field.
Background references:	See good practice references.
Good Practice:	***In briefing the design team and assigning responsibilities for environmental matters, clients and their consultants or design-and-build contractors should consider:*** • any established project environmental policy and its provisions for responsibilities and briefing the design team; • if no environmental policy has been established, setting down the key environmental features of the project in a succinct manner, for distribution to all team members; • reviewing the recommendations of this handbook and the BSRIA code of practice and applying those recommendations appropriately to their project procedures; • setting out defined responsibilities in a clear and unambiguous manner for distribution to all team members; • distributing widely the complete design brief or its environmental provisions; • establishing project environmental issues as a regular item on the agenda for design progress meetings; • holding additional team meetings specifically on environmental matters because such additional meetings will heighten team members' awareness and serve to focus attention on what may be new areas of concern for many team members; • holding seminars on key environmental issues if the team's knowledge warrants them.
Good practice references and further reading:	• *Design Briefing Manual*, AG1/90, BSRIA, 1990. • Barwise, J., and Battersby, S., *Environmental Training*, Croner Publications, 1993. • Halliday, S.P., *Environmental Code of Practice for Buildings and Their Services*, BSRIA, 1994. • Miller, S., *Going Green*, JT Design Build, Bristol, undated. • Construction Industry Environmental Forum, *Environmental management in the construction industry*, Notes of meeting held on 22/09/92, CIRIA, 1992.

Stage D2   Inception and feasibility

Stage D2: Inception and feasibility covers design inception and feasibility studies, examines a range of environmental constraints which a particular site may impose on a project and identifies the way to deal with them. The environmental assessment process is introduced with regard to legal requirements and the appropriate time to carry one out. Planning issues are only mentioned in respect of the timing of an outline application since the broader issues of land-use planning are to be covered in a separate volume. Good environmental practice involves incorporating environmental considerations into the design process from the outset.

D2.1   Legislation and policy

D2.2   Environmental assessment

D2.3   Site issues

D2.4   Securing outline planning permission and identifying financial implications

D2.1   Legislation and policy

Issue D2.1	Legislation and policy
Current legal position: ▸ D3.2.1 ▸ D3.2.2 ▸ D3.7.1	Many of the environmental issues and legal requirements relating to construction projects should be ascertained early in the design and specification phase. There are several areas of environmental legislation and regulations which will have to be considered both at domestic UK law and EC levels. An explanation of the coverage of UK law and in particular the differences between the law in England and Wales, in Scotland and in Northern Ireland is given in the Introduction on page 2. *Legislation* There is a considerable volume of legislation already in existence with the purpose of protecting the environment. There are many areas in which there may be future developments on the law and environmental policy. A distinction can be made between local environmental issues which arise during actual construction such as waste water and solid waste and the wider global environmental context. The legislation mainly takes two forms and these are often interrelated: • the requirement for environmental permits, consents and authorisations e.g. for discharges to water and air, consents governing noise, emissions, hazardous substances and planning permission – breach of these is usually a criminal offence; • the imposition of civil or criminal liability in respect of other matters e.g. the duty of care as respects waste (Section 34 Environmental Protection Act 1990), noise or vibration constituting either statutory nuisance (Sections 79-82 Environmental Protection Act 1990) or common law nuisance. These two elements will affect the inception of the project and its feasibility e.g. whether planning permission is granted or not and, if it is, the conditions attached to it such as the level of clean-up required of any contamination. The following is a only a brief summary of the areas of law which should be considered, many of which are dealt with in more detail in specific issue boxes. It should be remembered that no two construction projects are the same. The list below identifies the main areas and it is always necessary to explore the legal issues in each case:
▸ C2.1.2 ▸ C2.1.3 ▸ C2.1.4 ▸ C2.1.5 ▸ C2.1.9 ▸ D4.2.2	• wildlife and countryside law (see D3.2.2); • energy use and efficiency, global warming and climate change (see D3.7.1); • town and country planning law including the environmental assessment of development projects (see D3.2.1); • health and safety law; • law on the use of building materials (see D3.2.2) and hazardous substances (see D4.2.2); • law of waste and recycling (see C2.1.2 and C2.1.3); • pollution control law – emissions to air, water, soil (see C2.1.4, C2.1.5 and C2.1.9); • public access to information. *Common Law* It is also important to consider environmental liability which may arise at common law. This will largely arise where third parties have suffered damage to themselves and/or their property e.g. by contamination of their land or drinking water. The main areas of common law are: • **Nuisance** – the tort of nuisance is basically an act or omission on certain land which unreasonably interferes with or disturbs another person's right of enjoyment of other land. Nuisance may be either public nuisance or private nuisance. Public nuisance is a criminal offence as well as a civil wrong, and only applies if the defendant's act or omission is affecting a significant section of the public as a whole, for example where a contaminated site is polluting a drinking water supply. • **Negligence** – requires proof of fault, i.e. conduct falling below a standard that courts would regard as 'reasonable'. The plaintiff must also show both that he has suffered damage (and not mere economic loss) and that his damage has been caused by the negligence complained of. Additionally, the defendant must owe a duty of care to the particular plaintiff. In the case of a contamination of land, such a duty exists between owners and occupiers of neighbouring property. • **Rule in *Rylands –v– Fletcher*** – imposes strict liability for all damage resulting from a person having brought something on to his land that is not naturally there and which is likely

Issue D2.1	Legislation and policy
Current legal position continued:	to 'do mischief' if it escapes. This would include, for example, toxic chemicals, and anything that might cause purely physical damage such as water held back by a dam. It is no defence to prove that all possible precautions have been taken to prevent damage resulting from an escape. • **Trespass** – trespass to land occurs when there has been any unjustifiable intrusion by one person upon the land of another. Contaminants escaping from a contaminated site and entering upon the land of a neighbour would accordingly constitute a trespass.
References to the current legal position:	***The main UK Acts of Parliament are:*** • Alkali Works etc Regulation Act 1906. • Clean Air Act 1993. • Control of Pollution Act 1974. • Environmental Protection Act 1990. • Water Resources Act 1991. • Water Industry Act 1991. • Health and Safety at Work etc Act 1974. • Town and Country Planning Act 1990, Town and Country Planning (Scotland) Act 1972, Planning and Compensation Act 1991, and Planning (Consequential Provisions) Act 1990. • Planning (Hazardous Substances) Act 1990. • Building Act 1984, Building (Scotland) Act 1970, and particularly the 1991 Building Regulations. • Occupier's Liability Acts 1957 and 1984. • Noise and Statutory Nuisance Act 1993. ***Other legislation:*** • The numerous regulations under the above Acts of Parliament. • Environmental Information Regulations 1992 (SI 1992 No. 3240).
Policy and forthcoming legislation:	EC policy is broadly contained in the 5th Environmental Action Programme entitled *Towards Sustainability*. The programme is a policy document for environmental legislation to the year 2000. There are five target areas tourism, industry, energy, transport and agriculture. Current EC policy is a result of earlier Action Programmes. Policy documents may also focus on a specific area: for example, the EC has policies on waste and energy in addition to the 5th Environmental Action Programme. Current UK policy is contained in the white paper *This Common Inheritance* and the yearly reports on implementation of the proposals contained in that document. Policy focuses include aspects of climatic change and global warming, economic instruments and the introduction of market mechanisms to promote environmental protection, and energy efficiency. A paving bill is expected in the 1993/94 Session which will contain provisions for the establishment of an Environment Agency for England and Wales and a Scottish equivalent. The proposed Construction (Design and Management) Regulations – CONDAM – will apply to all projects which include construction work. The regulations set out responsibilities for developers, designers and contractors in respect of health and safety during construction.
Policy references	**EC** • Draft Directive on Landfill of Waste (Com (91) 102). • Fifth Action Programme on the Environment: *Towards Sustainability* (OJ C138 17.05.93), 1992. • Draft Proposals for a Directive on a tax on carbon dioxide emissions, energy efficiency and the promotion of renewable energy (Coms (92) 226, 182 and 180). • Green Paper on *Proposals for Remedying Environmental Damage*, 1993. **UK** • Government White Paper: *This Common Inheritance – Britain's Environmental Strategy (1990)* and *Yearly Reports* (1991 and 1992). • HSE Consultative document – *Proposals for Construction (Design and Management) Regulations and Approved Code of Practice*, 1992.

D2.2 Environmental assessment

Issue D2.2	Environmental assessment
Background:	An environmental assessment (EA) is a systematic analysis, using the best practicable techniques and sources of information, of the environmental effects of a civil engineering project or large scale building development. It is normally undertaken by or on behalf of the client and is taken into account by authorities before granting permissions. They have been subject to public scrutiny under the planning system for some time. However, in 1985, a European Directive (85/337/EEC) formalised EA procedure, emphasising the need for a systematic analysis of information and evaluation of predicted effects. This Directive was implemented in the UK through a number of statutory instruments including the Town and Country (Assessment of Environmental Effects) Regulations 1988. The report resulting from an EA is known as the Environmental Statement (ES) for the project. Schedule 1 of the regulations lists those projects where an EA is mandatory. Schedule 2 provides indicative criteria and thresholds to establish when an EA *may* be required for other types of project. Schedule 3 broadly defines the content of an EA. The Directive states that, where appropriate, an outline of the main alternatives studied by the client, together with an indication of the reasons for the choice may be presented and in a form which provide a focus for public scrutiny of the project. There can be benefits in carrying out an EA at the earliest possible moment, in some cases even at pre-acquisition stage. However, there are certain impacts which will not become clear until an outline scheme has been designed, so, in the majority of cases, the ES will accompany that of an outline planning application. In addition, an EA is likely to start with a initial scoping exercise which is used to identify those subject areas which need addressing in the main EA. For many civil works, for example motorway schemes, the process of assessing potential environmental effects is an integral part of the comparison of different scheme options. CIRIA SP96 provides an extensive introduction and guide to the identification, evaluation and mitigation of environmental impacts in development schemes covering construction and operational phases. It has separate sections on river and coastal engineering, water engineering, linear development, electricity generation, minerals extraction, waste management and building development. There are 11 stages in the process, the main ones being: • initial consideration of the development options; • identification of the key issues for the scheme – the scoping; • establishment of the environmental characteristics of the site – the baseline survey; • prediction and assessment of the environmental impacts; • identification and design of possible mitigation measures. The environmental impact of buildings during their lifetime is considered by the BRE Environmental Assessment Method (BREEAM). The basis of the scheme is a Certificate awarded to individual buildings stating clearly the performance of the building measured against a set of criteria for a range of issues. BRE appoints independent assessors and is also responsible for specifying the criteria and method of assessment used. BREEAM currently exist for *New Homes* (3/91), *New Superstores and Supermarkets* (2/91), *New Office* (1/93) and *Existing Offices* (4/93) and one for *Industrial Buildings* is due to be published soon. An EA carried out under the regulations and BREEAM differ in a number of ways, including: • EA is carried out as part of the project planning and design process, is enacted by legislation and is mandatory for certain projects, and relates to all major construction projects; • BREEAM relates to the design, maintenance and management of a building, is voluntary and, at present, relates only to certain types of building.
Background references:	• *Environmental Assessment: Guide to the identification, evaluation and mitigation of environmental issues in construction schemes*, CIRIA Special Publication 96, 1993. • *Environmental Assessment: A guide to the procedures*, DoE and Welsh Office, 1989. • *Environmental Assessment*, Department of the Environment Circular 15/88. • Department of Transport, *Manual of Environmental Assessment*, HMSO, July 1993. • BREEAM 1/93, *An environmental assessment for new offices*, BRE, 1993. • BREEAM 2/91, *An environmental assessment for new superstores and supermarkets*, BRE, 1991. • BREEAM 3/91, *An environmental assessment for new homes*, BRE, 1991. • BREEAM 4/93, *An environmental assessment for existing office buildings*, BRE, 1993.

Issue D2.2	Environmental assessment

Good Practice:	*When considering whether, when and how to undertake an environmental assessment for your project:*
	
	• read, review carefully and act on the DoE/Welsh office guidance published in *Environmental Assessment: A guide to the procedures*;
	• consider the guidance in CIRIA Special Publication 96 on *Environmental Assessment: Guide to the identification, evaluation and mitigation of environmental issues in construction schemes*;
	• liaise at a very early stage with the planning authority or other appropriate agency that will need to give consent or formal approval to the scheme;
	• recognise that the environmental statement should include the following components:
	  – a description of the civil engineering works or building development proposed;
	  – data identifying and assessing the effects of the works or development on the environment;
	  – a description of the likely environmental effects;
	  – a description of the measures to be taken to avoid, reduce or remedy any adverse effects;
	  – a summary in non-technical language;
	• review the pamphlet published by CPRE (Council for the Protection of Rural England) for planners, developers and environmentalists, which suggests three main areas where environmental statements frequently fall short of the ideal:
	  – identifying environmental impacts – scoping is identified as a key area including defining the parameters of the assessment and exploring alternatives, and policy context;
	  – assessing environmental impacts – methodology, in particular method and approach to assessment, and description and justification of forecasting methods;
	  – involving the public – public consultation including the early involvement of the public and free flow of information, and presentation including the provision of a non-technical summary of the information provided and adequate opportunities for debate;
	• recognise that there has been some criticism of the standard of statements already submitted (Tomlinson, 1990) particularly in three areas:
	  – the focus and content;
	  – the scientific and technical competence; and
	  – the organisational and presentational qualities;
	• recognise that an increasing number of civil engineering clients and building developers are voluntarily undertaking environmental assessment as an integral part of the design and planning of a their projects;
	▸ D3.8.3 • see D3.8.3 for details of BREEAM assessments.

Good practice references and further reading:	• *Environmental Assessment: A guide to the procedures*, DoE and Welsh Office, 1989. • Department of Transport, *Manual of Environmental Assessment*, HMSO, July 1993. • *Environmental Assessment: Guide to the identification, evaluation and mitigation of environmental issues in construction schemes*, CIRIA Special Publication 96, 1993. • Tomlinson, P., *Environmental Assessment and Management*, Chapter D1 in Curwell, S.R., March, C.G. and Venables, R.K., (Eds), *Buildings and Health*, RIBA Publications, 1990. • *Environmental statements: Getting them right*, CPRE, May 1990 • Therivel, R. et al, *Strategic environmental assessment*, Earthscan Publications, London, 1992. • Contact the Institute of Environmental Assessment on 0790 763613.
Legal references:	**EC** • Directive on the assessment of the effects of certain public and private projects on the environment 85/337/EEC. **UK** • Town and Country (Assessment of Environmental Effects) Regulations 1988 (SI 1988 No. 1199) (as amended by SI 1990 No. 367 and SI 1992 No. 1494). • Land Drainage Improvement Works (Assessment of Environmental Effects) Regulations 1988 (SI 1988 No. 1217). • Harbour Works (Assessment of Environmental Effects) Regulations 1988 (SI 1988 No. 1336) and (No.2) Regulations 1989 (SI 1989 No. 524). • Highways (Assessment of Environmental Effects) Regulations 1988 (SI 1988 No. 1241). • Electricity and Pipe-line Works (Assessment of Environmental Effects) Regulations 1989 (SI 1989 No. 167).

D2.3   Site Issues

Issue D2.3.1	Derelict and contaminated land

Background: ▸ D2.3.6 ▸ D4.2.3 ▸ D4.2.4	Contaminated land has been described as 'land containing substances that, when present in sufficient quantities or concentrations, are likely to cause harm to man, the environment or materials used in construction' (Viney & Rees 1990). Failure to take account of potential hazards may put at risk not only the workforce directly involved in redevelopment but also the people who will live or use the site thereafter. In most cases contamination is the result of previous land use, usually for some industrial purpose, or the result of inappropriately managed landfill which can produce considerable quantities of methane and leachates. In addition, there are instances where natural contamination is a problem, for example methane gas or radon emissions and high sulphate concentrations which can be detrimental to building materials. Risks to the surrounding environment may also arise from contaminant migration. Evidence presented to the House of Commons Environment Committee (1990) suggests that contaminated land may occupy as much as 50,000 hectares of the UK. DoE published figures put the total at 27,000 hectares. There are many legislative and regulatory instruments which have a bearing on contaminated land including the Town and Country Planning Act 1990. A series of guidance notes has also been produced by the Interdepartmental Committee on the Redevelopment of Contaminated Land established in 1976 by the UK Government.
Background references:	• Viney, I., and Rees, J., *Contaminated Land – Risks to Health and Building Integrity*, in Curwell, S.R., March, C.G. and Venables, R.K., (Eds), *Buildings and Health, The Rosehaugh Guide to the Design, Construction, Use and Management of Buildings*, RIBA Publications, 1990. • *Environmental issues in construction – A review of issues and initiatives relevant to the building, construction and related industries*, CIRIA Special Publications 93 and 94, 1993. • Construction Industry Environmental Forum, *Contaminated land*, Notes of a meeting held on 12/1/93, CIRIA, 1993. • Construction Industry Environmental Forum, *Contaminated land*, Notes of a meeting held on 14/9/93, CIRIA, 1993.

Issue D2.3.1	Derelict and contaminated land

Good Practice:	*When considering a project on a derelict or contaminated site, consider or recognise:*
	
	• that a systematic approach is needed, (see Viney & Rees):
	  – review data on the site's history;
	  – design and undertake an appropriate site investigation to determine whether a potential contamination problem exists and to assess any hazards;
	  – select the most appropriate use for the site and decide what, if any, remedial action is required;
	• the guidance available in CIRIA Special Publication 78 and other forthcoming CIRIA publications;
	• that British Standard Draft for Development DD175 (BSI 1988) gives guidance on techniques for the investigation of contaminated land;
	• consider the need to use expert advice in identifying the extent and nature of the problem and proposing mitigation measures;
	• that the objective in undertaking mitigation measures is to eliminate or at least reduce to an acceptable level, risks to public health, the environment and the fabric of a construction project;
	• that there are three main strategies to achieve this objective:
	  – removal to a licensed tip
	  – encapsulation or fixation
	  – on-site treatment and destruction of contaminants by chemical, biological or physical means,
	  the first two methods being the most used in the UK, although the third tackles the problem head on rather than transferring it elsewhere or trying to disguise it;
	• that government grant aid may be available for treatment and redevelopment of contaminated land, for example Derelict Land Grant and, in some cases, Supplementary Credit Approval.

Good practice references and further reading:	• As background references, plus: • Leach, B.A., and Goodger, H.K., *Building on derelict land*, CIRIA SP78, 1991. • Steeds, J.E., Shepherd, E. and Barry, D.L., *A guide to safe working practices for contaminated sites*, Unpublished CIRIA Core Programme Funders Report, FR/CP/9, July 1993, (in preparation as an open publication). • *Remedial treatment of contaminated land*, in 12 volumes, forthcoming CIRIA publications due to be published early 1994. • *Guidance on the sale and transfer of contaminated land*, Draft for open consultation, CIRIA, October 1993.
Legal references:	See D4.2.8.

Issue D2.3.2	Site hydrology and water quality

Background:	Population increase and rising per-capita demand have highlighted the need to protect and conserve the UK's limited water resources. In addition, wildlife habitats, especially wetlands, have been damaged or destroyed through water abstraction and land use change. With the establishment of the NRA, England and Wales now have an effective force for ensuring the maintenance of water quality and resources.
	The development of a site may adversely affect a local water resource in a range of ways:
	• diversion, temporary or otherwise, of a watercourse; • alteration of the local hydrology, for example by increased drainage, which may adversely affect the water table; • pollution of ground water or aquifer both during construction phase and after; • depletion of a local water resource through abstraction; • prevention of repletion of groundwater, for example by covering a site in hard surfacing and discharging run off via the sewerage system; • release of contaminants into water course as a result of site disturbance; • loss of wildlife habitat or landscape character.
	The National Rivers Authority is developing local catchment management plan in all its regions.
	Under the Water Resources Act 1991 it is an offence to cause or knowingly permit any polluting matter to enter controlled waters. The discharge of dangerous substances to controlled water is governed by discharge consents issued by the NRA or, if on the 'red list' of more polluting substances, by Her Majesty's Inspectorate of Pollution.
	Development in river flood plains has in the past caused flooding problems in times when a river channel's capacity has been exceeded. Whilst there are various engineering solutions available to prevent this from happening, more and more emphasis is being put on employing natural solutions where possible, including the creation of flood storage lakes and avoidance of building on areas susceptible to flooding.
	With coastal sites, the characteristics of the sea bed, of tidal conditions and the land/water interface will need to be given special consideration. In estuaries, an understanding of the dynamics of sediment deposition in relation to issues such as tidal regimes will be critical in terms of long term environmental assessment.
Background references:	• *A Manual for the Assessment of Major Development Proposals*, HMSO, 1981: Technical Note 1 – *Hydrological Considerations in the Appraisal of Major development*, and Technical Note 9 – *Water Pollution Considerations in the Assessment of Major Development*. • *Scope for control of urban run-off*, CIRIA Reports 123 and 124, 1992. • *Policy and practice for the protection of groundwater*, National Rivers Authority. • *A New Balance*: Jones Lang Wootton, McKenna and Co., Gardiner and Theobald, 1991 (p. 11).

Issue D2.3.2	Site hydrology and water quality
Good Practice:	A thorough understanding of the hydrology of a site is needed to avoid any adverse construction impacts or to foresee any opportunities for creating new wetlands and enhancing the overall water resource. ***Points for consideration include:*** • are there any local watercourses of ecological significance? • are there opportunities to make positive features out of site requirements, for example the creation of new ponds? • are there any potential drainage or flooding problems? • how will surface run off be disposed of, both during and after construction? • will the built development result in discharges of process water or effluent? • will discharge consents be required from the NRA? • will demolition and site clearance disturb potentially contaminative materials? • the implications of development on the local sewerage system, particularly if any effluent discharge or storm overflow into a watercourse is envisaged. In dealing with river or coastal engineering works, computer, mathematical, hydraulic or physical modelling may be appropriate. However, such work is often expensive and does not always answer environmental questions. Early consultation with the NRA and water services companies will be essential. A combination of licences and consents may be necessary: • discharge consent; • abstraction licence; • impounding licence; • land drainage consent. It may also be necessary to take account of the NRA's requirements as set out in the local catchment management plan.
Good practice references and further reading:	• *Environmental Assessment: Guide to the identification, evaluation and mitigation of environmental issues in construction schemes*, CIRIA Special Publication 96, 1993. • *A Manual for the Assessment of Major Development Proposals*, HMSO, 1981: Technical Note 1, *Hydrological Considerations in the Appraisal of Major Development* and Technical Note 9, *Water Pollution Considerations in the Assessment of Major Development*. • *Design of flood storage reservoirs*, Book 13, CIRIA/Butterworth-Heinemann, 1993. • Newbold, C., Purseglove, J., and Holmes, N., *Nature Conservation and River Engineering*, Nature Conservancy Council, 1983. • Lewis, G., and Williams, G., *Rivers and Wildlife Handbook*, RSPB/RSNC, 1984 (out of print – second edition due to be published in mid-1994). • *Ecodesign*, the Journal of the Ecological Design Association, Vol 1 No 4, Special Issue on Water, including article on the use of reedbeds to clean waste water.
Legal references:	• Water Resources Act 1991.

Issue D2.3.3	Flora, fauna and landscape
Background: ▸ D3.2.3 ▸ D3.3.2 ▸ D3.2.2 ▸ D4.5.1 ▸ D4.5.2	The majority of wildlife habitats in Britain have diminished greatly in size since World War II. At the same time support for and interest in nature and landscape conservation has grown enormously. The Royal Society for the Protection of Birds (RSPB) alone boasts almost 1 million members. The most important wildlife areas in the country are designated as Sites of Special Scientific Interest (SSSIs) by the Government's statutory nature conservation advisers English Nature (or their equivalents the Countryside Council for Wales or Scottish Natural Heritage), under provisions contained in the Wildlife and Countryside Act 1981. At the County level those sites which do not quite merit SSSI status, and thus are not given the same degree of statutory protection, are often designated in local plans as Sites of Nature Conservation Importance (SNCI) or similar. Marine Nature Reserves give statutory protection to wildlife sites below the low water mark. On a broader scale the European Community has adopted a Habitats Directive which will seek to protect important habitats and species. The Directive will require member states to designate Special Areas of Conservation (SAC) which will command protective measures. Habitats and species to be protected are listed in Annex I and II of the Directive. Other international obligations include the designation of Special Protection Areas (SPA) under the European Community Directive on the Conservation of Wild Birds (79/409/EEC), and the designation of Wetlands of International Importance under the Ramsar Convention. From time to time the Government issues Circulars and Planning Policy Guidance (PPG) on nature conservation, the most recent of which is Circular 1/92. A PPG drawing together all relevant advice on nature conservation is due to be issued in due course. In respect of landscape, the National Parks and Access to the Countryside Act 1949 established 10 National Parks in England and Wales. In 1988 the Government created a statutory Broads Authority which gave the Norfolk Broads the same status as a National Park and is currently considering establishing a similar body for the New Forest. Also designated under the 1949 Act are Areas of Outstanding Natural Beauty (AONBs) of which there are 38 in England and Wales covering 20,000 square kilometres. The Countryside Commission (Countryside Council for Wales or Scottish Natural Heritage) is the statutory agency responsible for advising the Government on countryside and landscape issues. Green Belt designation provides one of the most effective means of protecting land. Instituted as a planned tool to check urban sprawl and provide breathing space for urban dwellers, this land by no means necessarily comprises land of high intrinsic wildlife or landscape value. The National Rivers Authority has statutory duties in relation to the conservation and enhancement of the aquatic environment.
Background references:	• Hatton, C., *The Habitats Directive: Time for Action.* WWF UK, Surrey, 1992. • *Guidelines for the Selection of Biological SSSIs*, Nature Conservancy Council, Peterborough, 1989. • *River Corridor Surveys: Methods and Procedures*, Conservation Technical Handbook No. 1, National Rivers Authority, Bristol, 1992.
Good Practice:	*In protecting flora, fauna and landscape on and around building and civil engineering projects:* • ideally avoid any site of acknowledged high wildlife and/or landscape value, not least because attempts to develop such land will encounter resistance and delays; • consult at an early stage local authority planning departments which should be able to supply maps which identify SSSIs, National Parks, Green Belt and AONBs and possibly SNCIs or their equivalent; • consult Regional Offices of English Nature and the Countryside Commission (or their equivalents in other parts of the UK) which will be able to provide advice on particular sites;

Issue D2.3.3	**Flora, fauna and landscape**

Good Practice continued:	• consult the many voluntary sector organisations that take an interest in various aspects of nature and landscape conservation, in particular the Royal Society for Nature Conservation (RSNC) and the Royal Society for the Protection of Birds (RSPB) which are two of the most important in respect of wildlife; • recognise that individual sites may have some significant wildlife interest despite not being identified in local authority plans; • recognise that local people often have very strong feelings about their local wildlife and landscape and the absence of any official designation of a site does not automatically mean that there will be no wildlife or landscape interest; • consult the Countryside Commission and the Council for the Protection of Rural England (CPRE), and their equivalents in Wales and Scotland, on landscape issues; • during any initial scoping exercise of the site, identify potential impacts on local ecology or landscape; • consider employing consultants to undertake a full survey(s) and assessment(s) of the site, possibly at different seasons, to investigate the possible adverse impacts and the potential for mitigation; • recognise that this last point is particularly true in urban areas where any local green space of whatever condition may be deemed to have some value for wildlife; • remember that a development can often have an impact beyond the construction site, for example by causing damage to drainage, polluting air and water, by severing recognised wildlife corridors or, by the nature of the development's operational use, increasing pressure on vulnerable habitats or species; • recognise that, in some instances, the presence of certain species, for example nesting birds, may constrain when work can be carried out on a site; • utilise the available guidance listed here; • recognise that ecological value of a site is a consideration taken into account in BREEAM *New Offices* 1/93 in which credits are given for minimising ecological damage and enhancing existing interest; • use the simple checklist included in BREEAM to assess whether or not a site is of ecological value.
▶ D3.6	In outline design, also actively consider the positive benefits of landscaping at the feasibility stage.
Good practice references and further reading:	• *Environmental Assessment: Guide to the identification, evaluation and mitigation of environmental issues in construction schemes*, CIRIA Special Publication 96, 1993. • Construction Industry Environmental Forum, *Nature conservation issues in building and construction*, Notes of a meeting held on 23/3/93, CIRIA, 1993. • *Who's Who in the Environment 1992*, The Environment Council, London. A very useful directory of all statutory and voluntary sector environmental organisations. Separate editions produced for England, Wales and Scotland. • Prior, J.J., (Editor), BREEAM 1/93, *An environmental assessment for new offices*, BRE, 1993.
Legal references:	**EC** • Directive on the Conservation of Wild Birds (79/409/EEC). • Directive on the Conservation of Wild Fauna and Flora (92/43/EEC) (Habitats Directive) (Intended to be implemented by 05.06.1994). **UK** • Government White Paper: *This Common Inheritance – Britain's Environmental Strategy*, 1990 and *Yearly Reports* 1991 and 1992. • National Parks and Access to the Countryside Act 1949. • Wildlife and Countryside Act 1981 (as amended).

Issue D2.3.4	Landfill gas and naturally-occurring methane and CO_2

Background: ▶ D3.3.1	Landfill gas, i.e. gas that originates from landfill sites, is principally a mixture of methane and CO_2 derived from the anaerobic decomposition of organic matter. Other sources may be sewage sludge deposited on land; river and dockland silt, coal measures and peat deposits. Methane may also be present in the ground from fractured gas mains.
	Landfill gas may be lighter or heavier than air depending on the proportion of methane to carbon dioxide. It is able to migrate long distances through permeable ground or along fractures and services ducts, for example from old landfill sites to new housing estates. Concentration of methane gas in air in the range 5–15% (approximately) can lead to explosions in buildings; high carbon dioxide may cause death through asphyxiation, particularly where landfill sites have been capped or ground cover has low permeability.
	Domestic landfill sites over 30 years old are less likely to pose a threat to building works provided the decomposition processes have progressed effectively over that period.
Background references:	See good practice references.
Good Practice:	Construction projects should not normally be located on or adjacent to sites producing significant quantities of methane or carbon dioxide or containing significant quantities of putrescible material (Leach and Goodger 1991).
	To establish if there is a potential problem on your site, use the following guidance.
	• Consultation with local authority staff should reveal information regarding the history of a site and an indication of potential landfill gas contamination. If contamination is expected a thorough survey of the site should be carried out. This should include month-by-month sampling from a number of points to show:
	– the extent of area affected by gas;
	– the concentration of principal gases;
	– emission rates through a ground surface.
	• Less than 0.1% LEL (Lower Explosive Limit) presents a negligible risk. If it is up to 10%, then remedial measures will be necessary. Between 10 and 25% the risk is substantial and forced ventilation and alarm systems will be needed. If concentrations higher than 25% by volume are found, construction will not normally be permitted.
	If a problem has to be tackled, proceed as follows.
	• Techniques are available for the construction of safe works or buildings on gas-hazardous areas, although the risks and/or costs involved may make some projects financially too risky or the funding institutions reluctant to invest.
	• If high concentrations are found, then the following strategy should be adopted:
	• gas movement control by:
	– active extraction by pumping;
	– passive extraction by creating gravel filled trenches;
	– vertical barriers;
	– and/or horizontal barriers;
	• safe design of buildings, confined spaces, voids, manholes, services etc. including provision of a gas membrane across the ground floor area, sealing all vulnerable entry points and channelling gas into specially designed containment spaces where it can be safely dealt with by evacuation, dispersal and dilution into open air;
	• careful ongoing monitoring.

Issue D2.3.4	Landfill gas and naturally-occurring methane and CO_2
Good practice references and further reading:	• DoE Planning Circular 17/89, HMSO, 1989. • ICRCL 17/78, *Notes on the development and after-use of landfill sites*, ICRCL, 1990. • Leach, B.A., and Goodger, H.K., *Building on derelict land*, CIRIA SP78, 1991. • BRE, *Construction of new buildings on gas-contaminated land*, 1991. • *Methane and associated hazards to construction, a bibliography*, CIRIA SP79, 1992. • Hooker, P.J. and Bannon, M.P., *Methane: its occurrence and hazards in construction*, Report 130, CIRIA, London, 1993. • Crowhurst, D. and Manchester, S.J., *The measurement of methane and other gases from the ground*, Report 131, CIRIA, 1993. • Card, G.B., *Protecting development from methane*, Unpublished CIRIA Core Programme Funders Report, FR/CP/8, March (in preparation as an open publication). • Raybould, J.G., Rowan, S.P., and Barry, D.L., *Methane investigation strategies*, (in preparation for issue as CIRIA Core Programme Funders Report, FR/CP/14).

Issue D2.3.5	Non-ionizing radiation
Background: ▸ D4.2.6 ▸ D4.2.7	Non-ionizing radiation includes ultra violet, visible light, laser, infrared and radio frequency and microwave radiation. In general, the sources of these are connected with the use of certain types of equipment such as welding, or are natural phenomena. However, electromagnetic radiation can be given off by underground and overhead power cables and studies have been carried out to investigate any links between the incidence of cancer and electromagnetic radiation from high voltage overhead power cables. No evidence of a direct causal link has been discovered although there is concern that exposure may promote the effect of other carcinogens.
Background references:	• *A New Balance*, Jones Lang Wootton, McKenna & Co., Gardiner & Theobald, 1991, page 17.
Good Practice:	There is no recommended good practice in respect of site-related non-ionizing radiation. However, publicity given to the possible links between high voltage power cables and the incidence of cancer has caused concern. A site which has local high voltage power cables, particularly overhead, may be more difficult to market than one without.
Good practice references and further reading:	• Bodsworth, P., and Bradley, A., Ionizing and Non-ionizing Radiation in Construction, Appendix 3 in Curwell, S.R., March, C.G. and Venables, R.K., (Eds), *Buildings and Health, The Rosehaugh Guide to the Design, Construction, Use and Management of Buildings*, RIBA Publications, 1990. (pp.247–250).

Issue D2.3.6	Radon and thoron

Background:	Radon-222 is a natural radioactive gas formed by the radioactive decay of uranium. There are traces of uranium in all soils and rocks across the UK and consequently radon is continuously being emitted from the ground. It has a radioactive half-life of 3.8 days and is rapidly diluted when it reaches open air. However, the gas can enter and collect in buildings through gaps in the floor structure. Exposure to high concentrations of radon for long periods increases the risk of people developing lung cancer and may be responsible for the premature deaths of 2,000 to 3,000 people a year in the UK.
	Radon-220, also known as thoron, has a half-life of 1 minute. The radiation doses people receive from thoron are on average about one-tenth of those from Radon-222.
	Radon concentrations in a building are related to four factors:
	• concentration of radium in the ground – the highest concentrations of radon are associated with granite masses in South West England and parts of the Pennines and Scotland; • permeability of the ground; • penetrations of the building fabric; • preventing ingress by artificially reducing the pressure in a void under the building.
	Building materials, in particular those containing crushed granite and phosphogypsum, also emit small amounts of radon. The average radon concentrations for all UK dwellings is 20 becquerels per cubic metre of air. The activation level for radon control in new buildings is 200 becquerels per cubic metre of air and 400 becquerels per cubic metre for existing buildings.
Background references:	• Curwell, S.R., March, C.G. and Venables, R.K., (Eds), *Buildings and Health, The Rosehaugh Guide to the Design, Construction, Use and Management of Buildings*, RIBA Publications, 1990 • *A New Balance*: Jones Lang Wootton, McKenna and Co., Gardiner and Theobald, 1991 (pp 14-15). • see D4.2.4.
Good Practice:	*When considering a site where radon might be present:*
	• be aware that the Ionising Radiations Regulations 1985 (IRR), issued under the Health and Safety at Work etc. Act 1974, make it an offence for employers to expose their employees to radiation doses (including radon) above specified limits; • recognise that the IRR are complex and provide for different levels of radon monitoring and control for different levels of employee exposure;
▸ D3.2.1	• be aware that the Building Regulations 1991 for England and Wales require that 'precautions shall be taken to avoid danger to health caused by substances found on or under the ground to be covered by the building' which includes any substance which is or could become radioactive; • study the *Householder's Guide to Radon* recognising that most of Cornwall and parts of Devon are the areas most affected, with recent designations, based on continuing research, of parts of Derbyshire, Northamptonshire, Somerset, Deeside, and Helmsdale, Yorkshire;
▸ D4.2.3	• seek advice from the National Radiological Protection Board as to the likely risks;
▸ D4.2.4	• recognise that if doses above a certain level are predicted, the project client may need to appoint a Radiation Protection Adviser; • at the appropriate time, study the guidance provided in D4.2.3 and D4.2.4.
Good practice references and further reading:	• See Background references and D4.2.4, plus: • *Householder's Guide to Radon*, 3rd Edition, DoE, 1992. • *Radon: Guidance on protection measures for new dwellings*, Report BR211, BRE, 1992. • Health and Safety Commission, *Exposure to Radon: Approved Code of Practice –* Part 3, HMSO, 1988.
Legal references:	See D4.2.3.

Issue D2.3.7	Noise
Background:	Noise can have a very disruptive effect on the lives of people particularly if it has an unpleasant tone or character, or if it persists over long periods of time. It can result in physical tiredness, sleep disturbance, lack of concentration and, if loud enough, hearing damage. Noise impact in respect of a construction project needs to be considered in a number of ways, including: • ambient noise levels, created on the majority of sites by the wind and movement of trees, or generated by machinery, traffic or the actions of people; • noise created as part of the construction process, for example by construction traffic or machinery; • noise as a result of site after-use, for example a motorway or railway line or, at a more local level, a building housing machinery or a night club. Some noise-sensitive land uses would include dwellings, in particular rest and retirement homes, schools and hospitals. The Control of Pollution Act 1974 empowered local authorities to serve notices imposing requirements on the way in which construction, demolition and certain other specified works in relation to built development are to be carried out for the purpose of minimising noise. A number of other statutory controls in respect of the control of construction noise also exist.
Background references:	• *Environmental Assessment: Guide to the identification, evaluation and mitigation of environmental issues in construction schemes*, CIRIA Special Publication 96, 1993. • DoE Circular 10/73 – *Planning and Noise*. • DoE Circular 2/76 – Control of Pollution Act 1974, Implementation of Part III. • HSE, *Noise in Construction*, 1992.
Good Practice: ▸ D3.4.3 ▸ D4.6.3	*The key points on noise to consider at this stage include:* • the potential for the project to reduce noise impact for neighbours and users; • the need to actively control noise impact of the project on immediate neighbours and the local community, at the construction stage as well as after completion; • the potential noise impact on the project from neighbouring sources, for example on a hospital project near to a noisy industrial site; • the need for suitable mitigation measures – see D3.4.3 for more details – the options including noise barriers such as earth mounds, fences, hedges and lines of trees as well as specially-engineered fence-like barriers; • using natural features to attenuate noise; • noise control in the internal environment which is covered in D4.6.3.
Good practice references and further reading:	• See background references, plus: • DoE Circular 8/81, The Local Government Planning and Land Act 1980, various provisions. • DoE Draft Planning Policy Guidance (PPG 23), *Planning and Noise*, 1992. • DoE, *A guide to noise abatement zones*. • HSE, *Noise in Construction*, 1992. • BS8233: 1987 *Sound Insulation and Noise Reduction for Buildings*, British Standards Institution, Milton Keynes. • *Use of vegetation in civil engineering*, Book 10, CIRIA, 1990.
Legal references:	• See D2.1.5, plus: • Control of Pollution Act 1974. • Road Traffic Act 1974. • Health and Safety at Work etc Act 1974. • Road Traffic Regulation Act 1984. • Noise at Work Regulations 1989 (SI 1989 No. 1790). • Environmental Protection Act 1990. • Town and Country Planning Act 1990 and The Town and Country Planning (Scotland) Act 1972.

Issue D2.3.8	Services Infrastructure
Background:	If the site has been developed in the past and has been cleared of all previously existing structures and buildings there is every likelihood that there will be services eg. sewers and water mains, electricity cables etc. remaining underground. In a few cases these may be of such size and importance that to disturb them will cause unwelcome disruption to neighbours and may prove prohibitively expensive. This may in turn impose constraints on what type of development is suitable for the site and its precise location. There may also be a risk of ground contamination from leaking services such as oil storage tanks. On the other hand, building over disused services may cause little or no hazard.
Background references:	No specific references identified.
Good Practice: ▸ D2.3.1	• Early consultation with relevant authorities, public utilities, service providers and neighbours should enable identification of any existing services and the problems which they may impose on development. • Continuing consultation and the employment of appropriate expertise will assist in the identification of appropriate strategies for dealing with such problems. • The cost, difficulties and timescales of providing basic utilities for the site need to be identified. • Consultation about 'greenfield' sites that have not been conventionally developed before should take into account that there may be land drains which will need to be diverted around the works or development or re-instated after completion. • Consultation about previously-used sites should seek to identify past uses of the site.
Good practice references and further reading:	No specific references identified.

Issue D2.3.9	Local community
Background:	In developing any site, the immediate local community will have a legitimate interest and concern in the size, shape and form of what is to be constructed, the working practices to be employed and who it is intended to serve. They may be concerned in respect of a number of issues, of which property values and quality of life will almost certainly be of paramount importance. They may adopt a Not In My Back Yard (NIMBY) stance against development without any truly justifiable argument to back it up. On the other hand they may have legitimate, wide-ranging concerns and may prove potent adversaries to development. Conversely, if they feel that their views are being listened to and concerns are being met, they can be useful allies. 6000 to 9000 complaints are made to Environmental Health Officers each year about construction and demolition work, of which some 15% relate to noise issues. The other main categories of nuisance include dust and dirt, visual impact and uncertainty. Designers and clients should take great care to ensure that local communities are fully informed of plans and proposals irrespective of scale. There is understanding that good communication is cost-effective and early involvement of community representatives has the advantage that: • local knowledge can be brought into play; • concerns of all parties are more readily understood and solutions accepted; • as contact builds up between the relevant parties greater trust and confidence can result. Many designers actively involve intended end users from the earliest possible moment in formulating and refining design proposals. In building projects, this process is known as community architecture and is particularly popular in large social housing schemes.

Issue D2.3.9	Local community
Background continued:	When a scheme requires an environmental assessment, one of the purposes of producing an environmental statement is to help explain to the general public the likely effects of the project and to enable them to form their own judgement on the significance of environmental issues raised by the project. On many civil projects, for example new road schemes, the local community will be actively consulted about alternative layouts and routes.
Background references:	• Construction Industry Environmental Forum, *Environmental issues in construction – A review of issues and initiatives relevant to the building, construction and related industries*, CIRIA Special Publications 93 and 94, 1993. Chapter 7 – Planning, land use and conservation, 1992. (pp. 7.18, 7.20–21). • *Environmental Assessment: Guide to the identification, evaluation and mitigation of environmental issues in construction schemes*, CIRIA Special Publication 96, 1993.

Good Practice:
 	
▸ D5
	First, there is a need to identify community and pressure groups that may have a bearing on the project and special interest groups that may be able to assist.
	
	Secondly, *in deciding on the need or desirability of involving the local community and if so, how, consider:*
	
	• the size and type of community to be affected by the project and establish an understanding of their perception of the problems associated with it;
	• consulting the local authority which may be able to assist with identifying the extent of community affected and their understanding of the project;
	• undertaking some basic research or social survey,
	• holding a public consultation exercise including:
	  – public meetings;
	  – publication of information brochure about the project options;
	  – establishment of exhibitions and information centres;
	  – establishment of one point of contact for information;
	• the recommendations of the Environmental Law Federation that contractors should:
	  – write to local people likely to be affected by the opening of a construction site;
	  – give an estimated completion date, and explain why this rarely can be precise;
	  – make sure people know the likely hours and days of work;
	  – give a contact name, address and telephone number in case of enquiry;
	  – make sure that there is an informative notice on the site displaying the above information, to cover anyone who might not have received a letter about the works (including visitors and passers-by);
	• review the use of a considerate contractors scheme for the construction phase of the project.

Good practice references and further reading:	• Wates, N., and Knevitt, C., *Community Architecture*, Penguin Books, London, 1987. • Construction Industry Environmental Forum, *Considerate builders and contractors*, Notes of a meeting held on 9/3/93, CIRIA, 1993. • *Environmental Assessment: Guide to the identification, evaluation and mitigation of environmental issues in construction schemes*, CIRIA Special Publication 96, 1993.

Issue D2.3.10	Local climate – assess impact of the local climate on development
Background: ▸ D3.3.5 ▸ D3.3.6	Buildings and other structures in the UK are developed with an overall understanding of the climatic conditions they are likely to have to withstand and the levels of comfort and safety which their users may demand. Local climatic conditions may influence the precise form of development which should be undertaken on a specific site. Latitude, altitude and topography will have a bearing on local climatic conditions and particularly on issues such as the amount and periodicity of sunlight and rainfall. Exposed positions may be unsuitable for buildings unless given some protection with shelter belts. Conversely a building designed to maximise solar gain will be of little use if, particularly during the winter, it is overshadowed by other buildings or by natural features. In cities, tall structures in exposed positions are likely to have to be designed against severe wind loading. Noise or air pollution may have a greater effect on a development if it happens to be in the lee of their source rather than upwind.
Background references:	• Prior, J.J., (Editor), BREEAM 1/93, *An environmental assessment for new offices*, BRE, 1993 (pages 17–21 give background on local wind effects and overshadowing).
Good Practice: ▸ D3.3.5 ▸ D3.3.6	*In considering the potential influence of local climate on the project design:* • understand local climatic conditions including sun path, seasonal temperatures, local wind and rainfall patterns, seeking assistance from the Meteorological Office and the local authority; • consider some form of climatic modelling; • consider the nature of the site's microclimate and assess any imposition it places on the design; • determine if the microclimate can be improved, for example by planting shelter belts.
Good practice references and further reading:	• Littlefair, P.J., *Site layout planning for daylight and sunlight: a guide to good practice*: BRE, 1991. • *Climate and Site Development*, BRE Digest No. 350, Part 1: General Climate of UK, BRE February 1990. • *Climate and Site Development*, BRE Digest No. 350, Part 2: Influence of Microclimate, BRE March 1990. • *Climate and Site Development*, BRE Digest No. 350, Part 3: Improving the Microclimate through design, BRE April 1990. • Consult the Meteorological Office or commercial environmental data companies for information on rainfall, wind, sunshine and temperature ranges for the site.

Issue D2.3.11	Local transport infrastructure

Background:	One of the greatest concerns about any new development, whether it be a building or civil engineering project, is its effect on the local transport infrastructure. One result may be dramatic increases in road traffic which the existing road system is incapable of handling. Local concerns will include: • road safety; • noise; • severance; • vibration; • visual intrusion; • land use; • air pollution; • delays. The consequences may be more disruption for local people as roads are modified to cope with the demand and/or the generation of antagonism towards a project which was otherwise viewed sympathetically. Such issues will inevitably have an influence on the size and siting of a development and possibly even its planned end use. Increased use of public transport is far less damaging to the environment than the use of private cars and walking and cycling are clearly much more environmentally positive. For these and other reasons the use of cars in the future is likely to be discouraged when compared with public transport, for example by restrictions on amount of car parking space associated with workplace developments, carbon taxes on fuel, tolls on highways and road pricing in cities. Large areas of car parking can be extremely wasteful of space and are often unattractive features, detracting from a well designed building or the surrounding landscape or townscape. In some cases it could be better to use the space for providing more building area or for enhancing its setting, but this may not succeed in discouraging car use and lead to parking on, and destruction of, the landscaping or on local residential roads. A balance needs to be struck.
Background references:	• Construction Industry Environmental Forum, *Environmental issues in construction – A review of issues and initiatives relevant to the building, construction and related industries*, CIRIA Special Publications 93 and 94, 1993. Chapter 7 – Planning, land use and conservation.
Good Practice:	*In striking a fair balance between the need for access to a completed construction project and the local transport infrastructure, consider:* • carrying out broad-ranging studies of the existing transport infrastructure and the potential impact of the proposed development upon it which, should be carried out by a team of specialist highway engineers and traffic consultants; • the proposed numbers of people using the completed development and how it is perceived they will reach it; • any possibilities for making good use of existing public transport links, for developing new ones which could be promoted as positive features of the project, or offering financial support; • the lessons to be learnt from other projects, where, for example, a developer of a large-scale housing development on the outskirts of London paid for the construction of a new station on an adjoining railway line; • if appropriate, introducing a new bus service, light rail system or water-based public transport; • limiting visitor car-parking spaces which may encourage people to travel by public transport and complete their journey by taxi or collection service; • if new roads are necessary, the incorporation of traffic-calming measures and special routes for heavy goods vehicle movements; • giving priority access to the site for pedestrians and cyclists; • careful linking of the project to existing footpaths and cycle routes; • careful positioning of car parks in relation to neighbours and noise-sensitive areas of the site.
Good practice references and further reading:	• Tolley, R., *Calming Traffic in Residential Areas*, Breti Press, Tregaron, Dyfed, 1990.

Issue D2.3.12	Cultural features

Background:	Cultural features could be described as ranging from earthworks, through historic buildings and landscapes to industrial heritage. As development and land use changes have slowly eroded wildlife and natural landscapes, so too have they had a similar effect on these cultural artefacts. There is a growing body of evidence that members of the public increasingly value cultural features in our landscapes.
	The Royal Commission on the Historical Monuments of England is responsible for recording historic monuments and structures in a national archive. It also co-ordinates the efforts of County Councils and Archaeological Trusts in preparing local sites and monuments' records. The Secretary of State for the Environment schedules monuments and Areas of Archaeological Importance under the Ancient Monuments and Archaeological Areas Act 1979 and a list of these is maintained by English Heritage. He also has a duty to prepare statutory lists of buildings of special architectural or historic interest and DoE Circular 8/87 is the most relevant document in this respect.
▸ D3.4.2	The Government requires planning authorities to have regard to the setting of buildings of special architectural or historic interest when considering proposals for new consultation which would affect them. Where a whole area is of special historic or architectural interest local authorities have powers to designate conservation areas. Sites close to conservation areas can also be affected by that proximity which can influence the development allowed.
Background references:	• DoE Circular 8/87, *Historic Buildings and Conservation Areas; Policy and Procedures*, HMSO, 1987. • Government White Paper: *This Common Inheritance – Britain's Environmental Strategy*, 1990. • *Environmental issues in construction – A review of issues and initiatives relevant to the building, construction and related industries*, CIRIA Special Publications 93 and 94, 1993. Chapter 7 Planning, land use and conservation.

Good Practice:	*In dealing with cultural features of a site, consider:*
	
	• avoiding disruption or development of sites with extensive, special or unique cultural features;
	• searching available records held by bodies such as the Royal Commission on the Historical Monuments of England, English Heritage and local authority planning departments;
	• consulting a county archaeologist, which may be able to offer consultancy, and/or a relevant local society;
	• consulting the Council of British Archaeology to track down appropriate national, county or local societies;
	• employing specialist consultants to advise on the relationship between the construction project and the cultural feature;
	• using the BPF Code of Practice on archaeology;
	• the potential impact of the project on cultural heritage, for example:
	
	   – land take;
	   – vibration and settlement;
	   – groundwater draw-down or rise;
	   – land use changes;
	   – damage caused by remedial measures (see CIRIA SP96 on *Environmental Assessment*);
	• investigating that impact, and consulting widely, at an early stage in the project;
	• consulting local Civic Trust, The National Trust or English Heritage, if appropriate;
	• consulting the NRA on any cultural features related to rivers, for example mills.
	
	Early investigation of the precise impact of a development on cultural heritage, along with full consultation with interested parties, is strongly recommended.

Good practice references and further reading:	• *Environmental Assessment: Guide to the identification, evaluation and mitigation of environmental issues in construction schemes*, CIRIA Special Publication 96, 1993. • British Property Federation, *Code of Practice on Archaeology*.

D2.4 Securing outline planning permission and identifying financial implications

Issue D2.4	Securing outline planning permission and identifying financial implications
Background: ▸ D3.2.1	Having completed a full assessment of the environmental factors influencing a particular site, now may be a good time in the project programme to consider making an application for outline planning permission and to assess the overall financial implications of the project. In particular, all constraints on the form, design and use of the works or development should have been identified and the views of the local community, planning authority and other stakeholders ascertained. The majority of development proposals will require local authority planning permission under the 1990 Town and Country Planning Act before being allowed to proceed. However, there are exceptions to this, for example where Enterprise Zones have been established, where a private or hybrid bill is promoted through Parliament, or where a trunk road is being promoted by the Department of Transport. Where an applicant wishes to gain confirmation from a planning authority that the project is acceptable, it is normal, at least with major developments, to submit an outline application providing it is not related to a listed building. This outline application will include only enough information to establish the principles of the project, such as the location, size and use. On the basis of information received, the planning authority will either grant outline planning permission, with or without certain conditions attached, or refuse. If the principle of development is accepted, then detailed designs can be worked up in preparation for a more detailed application. See D3.2.1 for guidance on appeals should an application be turned down.
Background references:	• Construction Industry Environmental Forum, *Environmental issues in construction – A review of issues and initiatives relevant to the building, construction and related industries*, CIRIA Special Publications 93 and 94, 1993. Chapter 7 – Planning, land use and conservation. • Davies, H.W.E., *Planning Control in Western Europe*, Department of the Environment, HMSO, London, 1989. Section E1 England.
Good Practice:	*Good practice in respect of seeking planning permission for a development and assessing its financial implications involves:* • making sure that all relevant environmental and other factors have been thoroughly investigated with a full Environmental Assessment if appropriate; • making sure the development proposals have been amended in response to these factors; • ensuring all relevant parties are consulted.
Good practice references and further reading:	• *Planning Applications – The RMJM Guide*, Robert Matthew Johnson-Marshall Ltd.
Legal references:	See D3.2.1.

Stage D3 Primary design choices and scheme design

Stage 3: Primary design choices and scheme design covers primary design choices and scheme design. It develops the client's requirements through outline design proposals to an approved scheme design, including application for full planning permission or other consents that may be necessary, particularly for civil engineering projects. It includes preparation of information for an approximation of construction costs and a subsequent cost estimate.

It involves selecting the form and configuration of the scheme which minimises impacts on the locality together with the selection of materials. Where the scheme involves buildings, energy, ventilation and lighting options need to be considered.

The stage also involves consultation with various bodies, including Department of Transport, Department of the Environment, National Rivers Authority, planning authorities, building control if appropriate, fire authorities, environmental authorities, utilities, licensing authorities and statutory undertakers.

D3.1   Briefing the design team

D3.2   Legislation and policy

D3.3   Site layouts and environmental impacts

D3.4   Impacts on occupants and users, visitors and neighbours

D3.5   Detailed consultation with relevant bodies

D3.6   Designing appropriate landscaping

D3.7   Energy options

D3.8   Labelling and other environmental information

D3.9   Ventilation options in buildings

D3.10   Daylight and need for artificial light

D3.11   Occupant comfort

D3.12   Criteria for primary material selection

D3.13   Use of materials

D3.1 Briefing the design team

Issue D3.1	Briefing the design team
Background: ▶ D1.4	The preparation of primary designs should build on the environmental principles established for the project with the relevant authorities through outline planning permission or other appropriate consent. The primary designs will set the overall tone of the project and comprise the first stage in realising what up until now will have been just broad concepts. As design progresses it becomes increasingly more difficult and more expensive to make radical changes. It is crucial therefore that, at this stage, all members of the design team are aware of, and understand, the implications of the project's environmental policy. They should be fully conversant with any environmental constraints or opportunities revealed through any informal or formal environmental assessment of the project, any site investigations and consultation, and be in agreement with the client regarding what can be achieved.
Background references:	• BS7750: 1992 *Specification for Environmental Management Systems*, British Standards Institution, Milton Keynes.
Good Practice: ▶ D2.2	*When briefing the design team at this stage:* • use and make available the outputs from any environmental assessments (EAs) undertaken; • use and make available site investigation (SI) reports and other relevant documentation; • consider organising formal seminars as training sessions, led by the consultants involved in the EAs and SIs, to ensure design team members fully understand the constraints and opportunities identified; • seek a commitment to team working; • consider establishing a schedule of regular design team workshops; • appoint a key member of the design team to be responsible for coordinating all environmental aspects of the design process.
Good practice references and further reading:	• Barwise, J., and Battersby, S., *Environmental Training*, Croner Publications, 1993. • *Design Briefing Manual*, AG1/90, BSRIA, 1990.

D3.2 Legislation and policy

Issue D3.2.1	Planning legislation
Current legal position: ▸ D2.2	The town and country planning legislation has created a sophisticated system of land use control and is a vital element of much environmental protection. The legislation controls the use of land in addition to the design and form of the built environment. In 1990, the planning legislation was consolidated. The principal acts (known as the Planning Acts) are now: • the Town and Country Planning Act 1990; • the Town and Country Planning (Scotland) Act 1972; • the Planning (Listed Buildings and Conservation Areas) Act 1990; and • the Planning (Consequential Provisions) Act 1990. Elements of all these acts were amended by the Planning and Compensation Act 1991. Under the town and country planning system, each planning area has a Development Plan. This sets planning policies and land use allocation for the area. These plans act as guides to future development but do not guarantee what is going to happen. • **Planning Permission** – The general rule is that, where development (as defined in the TCPA) is to be carried out, planning permission must be obtained. The General Development Order (SI 1988 No. 1813) grants 'deemed' planning permission for specified types of developments. The Use Classes Order (SI 1987 No. 764) further specifies 16 classes of activity where changes of use within a class do not constitute development and therefore do not require planning permission. In either case the relevant provisions may be restricted by conditions or planning agreements. It is true to say that almost all development undertaken by the construction industry will require express planning permission, although certain classes of development may be given deemed permission in enterprise zones or simplified planning zones. In general, there is a presumption in favour of development provided it does not cause demonstrable harm to interests of acknowledged importance. The Development Plan is important as its provisions will only be disregarded or over-ridden if 'material considerations indicate otherwise'. • **Procedure** – In order to obtain planning permission, an application is made to the local planning authority where the land is situated. The authority has a discretion to refuse planning permission or to grant it subject to conditions or, in certain cases, subject to a planning obligation under the TCPA. The legal requirement for applications and their decisions are found in the town and country planning legislation, and a host of secondary legislation and central government guidance (see Policy and forthcoming legislation below). An applicant is required to serve notice of the application on owners of land and any agricultural tenants. The local planning authority is required to give publicity to applications. This allows the public to make representations to the local planning authority prior to a decision being made on the application. • **Appeals** – An applicant may appeal against a refusal of planning permission to the Secretary of State for the Environment. An appeal may also be made if no decision is made by the local planning authority within eight weeks after a valid application for planning permission (deemed refusal) or sixteen weeks if accompanied by an environmental statement. An appeal must be made within six months of the refusal or deemed refusal. • **Enforcement** – There are several enforcement provisions contained in the planning acts. The most important are: – *Breach of Condition Notice (BCN)* – Where there is a breach of a planning condition the local planning authority may serve a BCN requesting compliance. There is no right of appeal against the service of a notice, although defences exist, and there are criminal sanctions for non-compliance. – *Planning Contravention Notice* – A local planning authority may issue such a notice requesting information on the use, operations or activities carried out on land where there is a suspected breach of planning control. Failure to comply is a criminal offence. – *Enforcement Notice* – This type of notice may be served where there is either development without planning permission or failure to comply with the terms of a

Issue D3.2.1	Planning legislation

Current legal position continued: ▸ D2.2	planning permission. There is a right of appeal to the Secretary of State, which suspends the notice. It is an offence not to comply with an enforcement notice. – *Stop Notice* – Served in the same circumstances as an enforcement notice, requiring the immediate cessation of the activity or development cited. This is rarely used except in extreme circumstances as compensation must be paid if a local planning authority gets it wrong. – *Injunction* – This can provide an urgent remedy to deal with an actual or threatened breach of planning control. • **Environmental Assessment** – In certain cases an application for a planning permission will be required to be accompanied by an environmental statement prepared through the process of environmental assessment. Environmental assessment is a technique and process by which information about the environmental effects of a new development, and of alternative schemes, are collected by the developer and taken into account by the local planning authority in deciding whether or not to grant planning remission. Generally an environmental statement is required in the case of major civil engineering developments, for example an oil refinery, motorway, thermal power station or waste disposal installation.
References to the current legal position:	• Town and Country Planning Act 1990. • Town and Country Planning (Scotland) Act 1972 (as amended). • Planning (Listed Buildings and Conservation Areas) Act 1990. • Planning (Consequential Provisions) Act 1990. • Planning and Compensation Act 1991. • Town and Country Planning (Use Classes) Order 1987 (SI 1987 No. 764). • Town and Country Planning (General Development) Order 1988 (SI 1988 No. 1813) (as amended). • Town and Country Planning (Assessment of Environmental Effects) Regulations 1988 (SI 1988 No. 1199) (as amended by SI 1990 No. 367 and SI 1992 No. 1494).
Policy and forthcoming legislation:	Planning policy is made at national, regional and local level. It provides the framework for making planning decisions within the procedures for getting planning permission set by the planning legislation. In summary, planning policy follows a hierarchial structure as follows. • *National Planning Policy* – DoE Circulars, Planning Policy Guidance Notes, Mineral Policy Guidance Notes, Consultation Papers. • *Regional Planning Policy* – Issued in respect of certain regions; either by Government in the form of Regional Policy Guidance Notes, or non-statutory guidance issued by regional groups of local authorities or other bodies. • *Development Plans* – Outside metropolitan areas the county planning authority prepares a structure plan and the District Council a local plan which together form the Development Plan whilst, in London and the Metropolitan Areas, the unitary authority prepares a Unitary Development Plan (UDP) which fulfils the same function as the local plan. • *Waste and Minerals Local Plans* – Prepared by the county planning authority.
Policy references:	• Government White Paper – *This Common Inheritance – Britain's Environmental Strategy*, 1990. • Government White Paper – *This Common Inheritance – Britain's Environmental Strategy, Second Year Report*, 1992. • DoE Circulars. • Regional Policy Guidance Notes. • Planning Policy Guidance Notes. • Mineral Policy Guidance Notes. • Structure Plans/Unitary Development Plans. • Local Plans. • Waste and Minerals Local Plans.

Issue D3.2.2	Building Regulations
Current legal position:	The Building Act 1984 is the statutory framework for building control, although the Building Regulations 1991 contain the detailed requirements and procedures. The current regulations, which became effective on 1 June 1992, impose requirements for carrying out certain building operations including the erection of new buildings or the making of a material change of use.
	The main requirement is that building work must be carried out in accordance with the technical requirements set out in Schedule 1. Approved Documents give guidance on ways of meeting the requirements. There are four main technical requirements of an environmental nature.
	• *Precautions against harm to health caused by substances on or in ground to be covered by the building* – Approved Document C lists types of sites likely to contain contaminants and possible contaminants and actions.
	• *Precaution against the permeation of toxic fumes from insulating material in to any building occupied by people* – Approved Document D relates this requirement to risks to health of persons from formaldehyde fumes given off by urea formaldehyde foams. If these foams are to be used there must be a continuous barrier to minimise, as far as practicable, the passage of fumes to occupiable parts of the building. The Approved Document gives technical solutions to the requirement including suitability of walls for foam filling, and installation in accordance with BS5618.
	• *Adequate means of storage and access to solid waste* – Provisions to meet compliance with the requirement, contained in Approved Document H, include the capacity, design and siting issues of solid waste storage.
	• *Provision for the conservation of fuel and power in certain buildings* – Approved Document L includes guidance on the limitation of heat loss through building fabric, controls for space heating and hot water supply systems and insulation of hot water storage vessels, pipes and ducts.
	Resistance to the passage of sound (Part E), ventilation (Part F) and heat-producing appliances (Part J) could also be considered.
	In complying with the technical requirements there is a further general requirement which provides that 'proper materials' appropriate for the circumstances are used and they are used in a workmanlike manner. Proper materials include those approved under the EC Construction Products Directive and the UK regulations implementing that Directive or a British Standard or Agrément Certificate.
	Exemptions from the requirements are set out in Schedule 2 of the regulations which include buildings on construction sites which are intended to be used only during the course of those works. There are additional obligations to notify and supply plans to the local authority when building work is intended to be carried out, and notification of the commencement of building work.

Issue D3.2.2	Building Regulations
Current legal position continued:	There is also a large body of legislation which may include building control measures. Therefore developers must check the local legislation in force in the area of the construction site to ensure compliance. The regulations do not at present cover certain important environmental areas such as heating efficiency, artificial lighting, ventilation, insulation foams. Overall the generality of the requirements limits their impact in compelling the use of safer building materials.
References to the current legal position:	**EC** • Directive on Construction Products 89/109/EEC. • Construction Products Regulations 1991 (SI 1991 No. 1620). **UK** • Building Act 1984. • Building (Scotland) Act 1970. • Building Regulations 1991 (SI 1991 No 2768) (as amended by SI 1992 No 1180).
Policy and forthcoming legislation:	Each technical requirement outlined in Schedule 1 is supported by a guidance document produced by the DoE – the 'Approved Documents'. Of particular concern are Approved Documents C, D, H and L and that relating to Regulation 7 which relate to the requirements mentioned above. The DoE have consulted on proposals to amend Parts F and L of the Building Regulations and Approved Document L so as to strengthen the requirements for the conservation of fuel and power. This review is in response to the Government's commitment, made at the 1992 Rio Earth Summit, to reduce carbon dioxide emissions. The proposals include the provision of insulation of both doors and windows, draught stripping of doors and windows, improved controls on domestic heating systems and efficient lighting with adequate controls. The DoE also invited views on where air conditioning and mechanical ventilation systems, which are energy expensive, should be allowed, and is proposing a method of minimising the use of air conditioning. A floor area of 500m^2 is proposed as the lower limit to which the revised regulations should apply.
Policy references:	• DoE, *Approved Documents to the Building Regulations*, 1985, 1990 and 1992. • British Board of Agrément, *Supplement to the Approved Documents* (updated quarterly). • Building Regulation Consultancy Service, *Guidance Document*, 1992. • DoE, *Energy Conservation: Proposed Amendments to Part L of the Building Regulations 1991*, 1993.

Issue D3.2.3	The Wildlife and Countryside Act 1981

Current legal position:	The Wildlife and Countryside Act 1981 has two main parts. • **Species protection** – Part I deals with the protection of birds, animals, and plants. In general, in relation to birds and animals, the offences relate to the killing, injuring or taking of any wild bird or any scheduled wild animal. There are further offences in relation to birds, of being in possession of a wild bird or its eggs, destroying a nest whilst being built or taking or destroying eggs. In relation to animals it is an offence to be in possession of a scheduled animal live or dead or any part of one, or intentionally damaging, destroying or obstructing places of shelter or protection of animals. In respect of plants it is an offence for anyone other than the owner or occupier of land to intentionally uproot any wild plant. It is a further offence to intentionally pick uproot or destroy any of the specified rare wild plants. (The Act also prohibits the sale or advertisement of wild animals and listed plants.) • **Habitat protection** – Part II allows for habitat protection and nature conservation by the designation of sites which are of special scientific interest (SSSIs) by reason of the existence of any flora, fauna or geological or physio-graphical features. Notice of designation of an SSSI must be given to the local planning authority for the area where land is situated, to the owner or occupiers of that land and to the Secretary of State for the Environment. Representations may be made in respect of such notification. The Notice will specify the reasons why the land is of special interest and any operations that may damage that flora, fauna or those features. SSSIs enjoy special protection. If an owner or occupier intends to carry out a damaging operation on that land he must seek consent from English Nature (and/or its equivalents in others countries of the UK) or planning permission. The Act also allows for the designation of Nature Reserves and Marine Nature Reserves. It is through the Act that relevant international and EC obligations are met, for example, the designation of Special Protection Areas for Birds as required by the EC Directive on the Conservation of Wild Birds and the designation of wetlands under the Ramsar Convention.
References to the current legal position:	**International** • Convention on Wetlands of International Importance especially as Waterfowl Habitat (Ramsar Convention), 1971. **EC** • Directive on the Conservation of Wild Birds 79/409/EEC (as amended). • Directive on the Conservation of Wild Fauna and Flora 92/43/EEC (Habitats Directive) (intended to be implemented by 5 June 1994). **UK** • Wildlife and Countryside Act 1981 (as amended). • Natural Heritage Act 1991.
Policy and forthcoming legislation:	Government policy aims to integrate environmental and economic activity in rural areas, to conserve the landscape, habitats and species diversity and to give extra protection to special areas.
Policy references:	• Government White Paper: *This Common Inheritance – Britain's Environmental Strategy*, 1990. • Government White Paper: *This Common Inheritance – Britain's Environmental Strategy – Second Year Report*, 1992. • DoE, *Action for the Countryside*, 1992. • DoE Circular 27/87 – *Nature Conservation*. • DoE Circular 1/92 – *Planning Controls over Sites of Special Scientific Interest*. • DoE Draft Planning Policy Guidance (PPG 24) – *Nature and Conservation*, 1992. • DoE Planning Policy Guidance (PPG 7) – *The Countryside and the Rural Economy*, revised 1992.

D3.3 Site layouts and environmental impacts

Issue D3.3.1	Re-use of land
Background: ▸ D2.3.1 ▸ D2.3.2 ▸ D2.3.4 ▸ D2.3.8	Development pressures, coupled with the desire to preserve greenfield sites, leads to increasing use of derelict land from former industrial or agricultural uses, landfill sites and infill sites in existing cities and towns. Many such sites will have existing buildings on them. Such land may be contaminated or have been used for fly tipping. There will be existing drains and services under the land to contend with. Consideration needs to be given to clearing and disposing of debris and/or removing, cleaning or encapsulating contaminated soils as well as the possibilities of re-using existing buildings. Feasibility studies should already have been undertaken at Stage 2 to assess the effect of these constraints before starting the primary design process. The decontamination of a site affected by industrial pollutants can be costly both financially and in terms of energy use and will need to accord with current regulations.
Background references:	• Jones Lang Wootton, McKenna & Co, Gardiner & Theobald, *A New Balance*, Sections 3.3.3 and 4.2.2, 1991. • Environmental Protection Act 1990. • DoE Circular 21/87.
Good Practice:	*Each site will be different but the following points offer some guidance:* • at an early stage in the design, a survey of the site and the existing features is required and the site history should be established from old maps etc; • enquiries should be made of statutory undertakings, water and drainage authorities (including the NRA) about existing services and drains under the site; • the local Environmental Health Officer is a key person to confirm the presence of any of the listed contaminative uses such as former chemical works, printing works, electricity generating stations, old railway sidings, landfill disposal sites etc. • if a site is known to have had a contaminative use, the extent of contamination and proposals for remediation should be assessed at planning stage: remediation can involve removal of soil to at least a depth of 1 metre over affected areas; • Building Control Authorities (Local Authority or Approved Inspector) should be able to advise on precautions needed against gaseous contaminants or landfill gases; • where development is proposed within 250 metres of a landfill or where there are known gaseous contaminants, the waste disposal authority must be consulted; • old underground mines or workings may need investigation and a full soil site investigation of the site ground conditions should be carried out. Remedial action will depend upon what is discovered, the extent and type of contaminants and the end-use of the land. *Options available are*: • disposal of contaminants to controlled landfill; • encapsulation with adequate soil cover and in-ground barriers; • on- or off-site physical, biological or chemical treatment. The redevelopment of the former landfill site at Stockley Park in West London, which involved the moving of large quantities of waste and earth within the site, is a good example of redevelopment of a derelict wasteland.
Good practice references and further reading:	• BS5930 1981: *Code of Practice for Site Investigation*, BSI, Milton Keynes. • BS DD 175 1988: *Draft Code of Practice for the Identification of Potentially Contaminated Land and its investigation*, BSI, Milton Keynes. • ICRCL Guidance Note 17/78, *Notes on the Development and After-use of Landfill Sites*, 1990 Edition. • ICRCL Guidance Note 59/83, *Guidance on the Assessment and Redevelopment of Contaminated Land*. • *Construction of New Building on Gas Contaminated Land*, BRE Report 1991. • Leach, B.A., and Goodger, H.K., *Building on derelict land*, CIRIA Special Publication 78, 1991. • Parking Business (Article on Stockley Park), in *Architects' Journal*, 25 July 1990.

Issue D3.3.2	Ecological value of site

Background: ▸ D2.2 ▸ D2.3.3 ▸ D3.2.3	Nearly all greenfield sites can be said to have some ecological interest and even brown-field sites (ie those that have been previously developed) may have acquired interest since being cleared. Interesting areas may include rank grassland, a length of hedgerow, scattered scrub, a short stretch of river, or a river corridor. On the other hand, the site may be an area of important habitat such as heath, wetland or ancient woodland, or provide a habitat for some rare species, such as the great crested newt. Preliminary site investigations should have identified if a site is of special importance for wildlife. Even if it is not, there are good reasons for attempting to conserve as much wildlife value as possible. A fabric of wildlife habitats is necessary in both town and country to ensure nature reserves and SSSIs do not become isolated green spaces in an otherwise largely urban landscape. What is more, people derive pleasure from observing wildlife. In a survey of 3,000 property companies carried out in 1990, most wanted a natural environment when they relocated (Nicholson-Lord 1991). Wildlife habitats can also provide natural barriers to pollutants and an attractive landscape setting. Careful attention to the configuration of a scheme may minimise ecological damage, conserving existing features and retaining landscape context. In cases where ecological interest is minimal or even absent, management of existing habitats, or creation of new ones, as part of the overall development scheme can increase ecological interest. Development on or adjacent to an SSSI will almost certainly necessitate mitigation measures and possibly even a demand for off-site ecological benefit as compensation. Under the Town and Country Planning Act 1990, local planning authorities may place a Tree Preservation Order (TPO) on any tree, group of trees or woodland (but not a hedge) in order to conserve the amenity of an area. Once a TPO exists, trees cannot be felled or branches cut without the consent of the planning authority. Under the Forestry Act 1967 nobody may fell trees of a certain size or number without a felling licence granted by the Forestry Authority. A Bill is currently going through Parliament which seeks to protect hedges which are of particular value for landscape, historic or wildlife reasons.
Background references:	• Nicholson-Lord, D., Healing the Sick, in *The Independent on Sunday*, 10 November 1991; article on architects' response to environmental concern. • *Protected Trees: A guide to tree preservation procedures*, DoE/Welsh Office, 1993.
Good Practice:	***In preserving the ecological value of a civil engineering or building site:*** • position works or buildings within a site to minimise ecological damage; • where possible retain existing ecological and landscape features, particularly those features which are important in retaining a sense of place; • take active steps to protect hedgerows and trees in the site, integrating them in the project if at all possible, and take steps to identify trees subject to a TPO; • try to avoid tidying up ponds or streams simply to apparently improve their appearance – such action can lead to loss of interesting wildlife species or habitats; • actively consider the appointment of ecological consultants to advise on the treatment, management and creation of areas of ecological interest; • ensure appropriate management of existing features and creation of new ones to increase the overall ecological interest and to provide other environmental benefits; • where possible create links between separate wildlife areas within and outwith sites; • establish effective mitigation measures for sites on or adjacent to an SSSI and consider the need for off-site ecological benefits to be provided as compensation; • consult English Nature (or equivalent), the NRA and nature conservation organisations such as the local Wildlife Trust, the local office of RSPB and the local authority; • seek advice from the Countryside Commission, the CPRE and the National Trust on landscape issues (or equivalents).
Good practice references and further reading:	• *Guidelines for minimising impact on site ecology*, RNSC, 1992. • Clifford, S., Local Colour: Article in *Landscape Design Extra*, July 1992. • Hough, M., *City form and natural process*, Croom Helm, 1984 • Buckley, G.P., (Ed), *Biological habitat reconstruction*, Bellhaven Press, 1989. • Ministry of Agriculture, Fisheries and Food, *Environmental Procedures for Inland Flood Defence Works*, 1992.

Issue D3.3.3	Transport and access to site

Background:	The industrialisation of the UK separated home from work and this pattern continues in the separation of the built environment into large, functionally determined areas: residential estates, retail parks, business parks, leisure complexes. As a result, there has been an explosion in the use of the private car to get from one place to another. Private cars are also extraordinarily convenient and once people have experienced using them they are reluctant to use other forms of transport. The Government forecast that by the year 2025 road traffic will increase by between 83% and 142% compared with 1988.
	Nevertheless the present policy in response to the problems of road congestion and pollution produced by cars is to find ways of encouraging people to use them less. The construction of a built environment around a public transport system sufficiently attractive for people to prefer it to driving has much to commend it. Good bus, rail or tram services within urban areas and good bus and rail services between settlements would be a step in the right direction.
	Of all methods of travel, the least energy-intensive is on foot or by bicycle. If the journey from home to work or home to school is relatively short, then it becomes possible to use these means of movement, particularly if bicycle lanes are provided. Many of these choices are of course in people's own hands, but designers can do something to help.
	In 1993 the Government issued a Consultation Paper entitled *Planning Policy Guidance on Transport* which seeks, inter alia, to provide guidance to local authorities on the role of land-use planning in reducing the need to travel and encouraging use of means of transport other than the car. The provision of adequate public transport is likely to be determined by politicians at national and local levels and the designer should exert influence through the normal democratic processes.
Background references:	• Vale, B. and Vale, R., *Green Architecture – Design for a sustainable future*, Thames and Hudson, London 1991. • *Roads for Prosperity*, Department of Transport Cm 693, 1989, HMSO, London. • Consultation Paper: *Planning Policy Guidance on Transport*, PPG 13, Department of the Environment. April 1993. • DoE & DoT, *Reducing transport emissions through planning*, HMSO, April 1993.
Good Practice: ▸ D2.3.11 ▸ C3.7.1 ▸ C3.7.2	*In making design decisions about the interaction between your project and transport:* • consider the location of buildings within the site in relation to available public transport; • actively consider with the client, if appropriate, the possibility of introducing subsidised bus services to serve a new development for an initial period and relate this to the size of car park provided; • take active steps to make the project or development as accessible to pedestrians, cyclists and people with disabilities as it is to those using private cars; • consider a mixture of uses distributed throughout a large site which are likely to encourage a balance between different forms of movement within the site – incorporating small workshops, business premises, shops and schools in close proximity to residential developments will encourage walking or cycling in preference to car use; • in planning for the construction phase, consult sections C3.7.1 and C3.7.2 of the companion handbook on construction.
Good practice references and further reading:	• Contact Transport 2000 on 071-388 8386. • Venables, R.K et al, *Environmental Handbook for Building and Civil Engineering Projects, Volume 2: Construction Phase*, CIRIA Special Publication 98, 1993. See Sections D2.3.11, C3.7.1 and C3.7.2.

Issue D3.3.4	Local transport Infrastructure

Background:	To reduce the levels of carbon dioxide and other pollutants entering the air, a reconsideration of a number of travel needs may be necessary in the future, for example, between home and work, access to food and other resources. For the present, a mixture of different forms of transport, both public transport and private cars, and provision for cycling and walking, are essential ingredients of a good local infrastructure.
	A clear hierarchy of roads, with distributor roads designed to take through traffic away from residential neighbourhoods and areas will reduce the danger and nuisance from such traffic. Reducing traffic speeds in residential areas means that cars can more safely mix with people who are either walking or cycling, and may also reduce fuel use.
	The secondary road network, distributor roads and local roads are determined by the planning authority as part of the local plan. This will usually involve extensive consultation with the local community.
	The extent to which the scheme, once operational, will generate traffic, together with the nature and volume of that traffic can assume fundamental importance when looking at larger scale civil engineering or building schemes. It may also affect the necessary road pattern, links to public transport, and requirements for footpaths and cycle routes.
	Unfortunately there is no accepted good practice in this area and there are conflicting views and priorities between the requirements of, for instance, convenience for cars, accessibility of/for buses and other public transport, access for deliveries, collections, refuse disposal and other services, pedestrian safety and security, creating a sense of 'place'. Improvements in local infrastructure may also be needed when civil engineering works are constructed, for example to provide access for operational vehicles to a waste management site, or when larger retail or business developments are built.
Background references:	• DoE & DoT, *Reducing transport emissions through planning*, HMSO, April 1993.
Good Practice:	• The prime aim must be to strike an appropriate balance between accessibility to the works or development, individuals' desires for privacy, safety, security and a pleasant environment in which to work or live, and the commercial viability of the scheme.
	• *In assessing how the project affects and is affected by the local infrastructure, consider or recognise:*
	– providing new access to the site which would direct works traffic away from any surrounding residential areas;
	– agreeing routes and providing signs for heavy traffic such as refuse wagons proceeding to a landfill site;
	– designing local distributor roads in the form of 'boulevards' or 'avenues' lined with trees to screen buildings from traffic;
	– that such roads could form areas of local activity with buildings containing shops and local facilities and/or businesses as well as residential buildings;
	– that in residential areas through traffic should be discouraged but that recognisable 'streets' permeable for residents and visitors should be created;
	– use of traffic calming measures, such as speed restricting humps and road narrowings to reduce car speeds with pedestrian crossings along distributor roads;
	– the possibility of providing special links for buses;
	– that pedestrian priority can be introduced where roads intersect;
	– that cyclepaths and footpaths need to be well overlooked where they are away from streets for safety and security;
▸ C3.7.1	– provision for cycle parks and stores within any works, facilities or building development;
▸ C3.7.2	– in planning for the construction phase, consult sections C3.7.1 and C3.7.2 of the companion handbook on construction.
Good practice references and further reading:	• DoT *Design Bulletin 32, Residential Roads and Footpaths*, 2nd Edition, HMSO, 1992. • Town design that puts building before traffic, Article in *Architects' Journal*, 13/1/93. • Venables, R.K et al, *Environmental Handbook for Building and Civil Engineering Projects, Volume 2: Construction Phase*, CIRIA SP98, 1993; see D3.7.1 and D3.7.2.

Issue D3.3.5	Building microclimate
Background: ▸ D3.3.6 ▸ D4.5.2	Designers can effect a measure of control over the climate of outside spaces around their buildings – their microclimate – by paying special attention to the form and layout of buildings. Designing to improve microclimate can help to reduce energy costs for space heating and make the spaces around buildings more comfortable and more useful. Access to daylight and sunlight is one aspect of the microclimate around buildings. The other main element of microclimate is shelter, from the wind as well as from excessive solar heat gain in summer. A favourable external environment will generally improve user satisfaction and it may improve the value of a building or make it easier to let or sell. The approach is relevant to the design of individual buildings or complexes, whether for residential or business use, and to civil engineering schemes such as power stations where large, monolithic structures are involved, stadia, chimneys and cooling towers. Beyond this there may be scope for similar microclimate approaches between adjoining sites, or for the complementary use of publicly owned non-developable land, perhaps for shelter belt planting.
Background references:	• *Climate and Site Development*, BRE Digest No. 350, Part 1: *General Climate of UK*, BRE February 1990, and Part 2: *Influence of Microclimate*, BRE March 1990. • Hough, M., *City form and natural process*, Croom Helm, 1984.
Good Practice: ▸ D2.3.10 ▸ D3.3.6	Ideally, microclimate should be a factor in assessing sites for many types of building. It is rarely the case however that a site can be selected on the basis of suitable microclimate as other factors generally carry more weight in site selection. (For more information about site selection and implications for site layout of the local climate, see D2.3.10) and read this section in conjunction with D3.3.6 overleaf. Improving the microclimate of a site by paying attention to the following aspects of the design of the building, landscaping and engineering works at an early stage is particularly important. *Solar access* • Design the road layout to run within 15° of E–W wherever possible so that building facades can be within 15° of south-facing. • Allow wide, south-facing frontages. • Plant deciduous trees to the south of dwellings. • Choose building type and form to limit overshadowing. • Use light-coloured paving materials to reflect sunlight and daylight into buildings. *Wind control* • Choose the form and arrangement of the building or works to avoid downdraughts and shelter external spaces. • Avoid large facades facing the dominant or critical wind direction. • Choose stepped or set back forms for tall buildings or structures. • Consider avoiding flat-roofed buildings and large cubical forms. • Create courtyards for maximum shelter and orientate partly open courtyards for optimum shelter from the dominant or critical wind direction. • Limit the maximum length of blocks, and avoid long parallel rows of smooth-faced buildings. • Provide draught lobbies or revolving doors at building entrances. *Shelter from landscape features* • Consider the use of trees, bushes, mounding and banking and other permeable features to provide wind shelter – shelter belts give shelter for five times their own height downwind. • Consider smaller scale planting of trees and bushes to give local protection to buildings or open spaces. A good example of an integrated landscape design to create a favourable microclimate is the landscaping at the Energy Park, Milton Keynes.
Good practice references and further reading:	• *Climate and Site Development*, BRE Digest No. 350, Part 3: Improving the Microclimate through design, BRE, April 1990. • Johnston, J. and Newton J., *Building Green – A guide to using plants on roofs, walls and pavements*, London Ecology Unit, 1993.

Issue D3.3.6	Overshadowing and access to daylight and sunlight

Background:	People expect good levels of daylight in a wide variety of buildings. Access to daylight and sunlight help to make buildings more energy efficient by reducing the need for electric light and also by contributing winter solar gain. There may, however, be an additional cooling load at some times of year.
	The design of the external environment is important in ensuring there will be an adequate quantity and quality of natural light in buildings, ensuring that the height of obstructing structures neither makes daylighting impossible nor blocks sunlight for much of the year.
	There may be circumstances where these considerations assume a greater or lesser importance. For example, in an historic city centre, a higher degree of obstruction may be unavoidable, if new developments are to fit in with the height and proportions of existing buildings. With a building, new or existing, where natural light and solar gain are of special importance, less obstruction and more daylight and sunlight may be necessary.
	In addition, new HSE regulations on the workplace require lighting to workplaces to be, as far as reasonably practical, provided by natural means.
Background references:	• Lynes, J.A., *Principles of Natural Lighting*, Elsevier, 1968.
Good Practice: ▶ D3.8.3 ▶ D3.7.3 ▶ D3.10 ▶ D4.6.5	*For good daylighting within a building*, the following should be borne in mind when designing external layouts. Considerations of view, privacy, security, access, enclosure and glare from glazed facades may also have an impact on natural lighting in a layout. • In new development, check that no obstructing building subtends an angle of more than 25° to the horizontal at a height of 2 metres above ground level, measured in a vertical section perpendicular to the main faces of the building. • In designing a new development or extension to a building it is important to safeguard daylight to nearby existing buildings. The right to light is a right which an existing building may acquire over the land of another, either by legal agreement or if it has been enjoyed without interruption for at least 20 years. • A new building should be set at a reasonable distance away from boundaries with adjoining land to avoid overshadowing and to enable future nearby developments to have similar access to daylight. Clearly this will not apply to a boundary next to a windowless flank wall of a building, nor where future development will also present a flank wall without windows. *Sunlight as an amenity* is valued in non-domestic buildings, is expected in living spaces in homes, and is also important in external spaces. • In general, spaces within buildings will appear reasonably sunlit when at least one main window wall faces within 90° of due south and receives at least a quarter of annual probable sunlight hours. • Care should be taken to safeguard sunlight to existing dwellings and non-domestic buildings in designing a new development. • Good site layout planning for daylight and sunlight should include ensuring adequate sunlight in the spaces between buildings, as this has an important impact on the ambience of a development. • Consider reserving the sunniest parts of a site for gardens and sitting out areas, or for focal points for views from within the building. • Shadier areas can usefully be set aside for car parking. • Poor sunlighting on the ground can occur where the building has a long face within 13° of due north or creates an enclosed or partly enclosed space (such as a court) which is blocked off from the southern part of the sky.
Good practice references and further reading:	• BS8206: 1992 Part 2 *Code of Practice for Daylighting*, BSI, Milton Keynes. • Littlefair, P.J., *Site Layout Planning for Daylight and Sunlight – A Guide to Good Practice*, BRE Report BR209, 1991. • BREEAM – See D3.8.3.

Issue D3.3.7	Passive solar design and its effect on site layouts

Background: ▸ D3.7	Since the pioneering work by A E Morgan at the Wallasey School in the early 1960s in this country, it has been recognised that, as well as bringing warmth and amenity to building spaces inside and outside, the sun is also a source of energy. This energy can be tapped in good building design to reduce the consumption of conventional fuels. This can be a particular priority in designing the form, fabric and site layout of a building as well as in the choice of heating and cooling systems, resulting in passive solar design. In the United Kingdom at least, passive design tends to be more cost-effective than using active solar systems. (For more information about both passive and active systems see D3.7 Energy Options). Passive solar design is valuable for both housing and non-domestic buildings. In non-domestic buildings half of the savings would result from savings in lighting energy consumption. Site-related factors will influence the decision of whether to opt for passive solar design, as well as the needs of the client and the intended use of the building. It will be easier to gain the full benefits of passive solar design on a south-facing slope than on a sloping site which faces north. High densities of development make it difficult to avoid obstruction or poor orientation on at least some part of the development.
Background references:	• Banham, R., *The architecture of the well-tempered environment*, The Architectural Press, London, 1969, pp 280-284. • Szokolay, S.V., *Solar Energy and Building*, Second Edition, The Architectural Press, London, 1978, pp 85–86. • Vale, B. and Vale, R., *Green Architecture – Design for a sustainable future*, Thames and Hudson, London, 1991, pp 71–72. • Lebens, R., *Passive Solar Heating Design*, Applied Science, London, 1980. • Turrent, D., Doggart, J., and Ferraro, R., *Passive Solar Housing in the UK*, Energy Conscious Design, London, 1981. • Mazria, E., *The Passive Solar Energy Book*, Rodale Press, 1979.
Good Practice:	Good practice in passive solar site layout design can be divided into the two key areas of orientation and overshadowing. *Orientation* • The main solar-collecting facades of domestic buildings should face within 30° of due south. • In houses the solar gain will be used most effectively if living rooms are on the south side with kitchens, bathrooms and garages to the north. • In non-domestic buildings, store rooms, computer rooms, canteens, and other rooms with high internal gains should be on the north side. *Overshadowing* • Care must be taken to ensure overshadowing by other buildings does not obstruct access to low angle sun in winter. • Tree locations are important. Deciduous species are best because they provide some shade in summer, but are leafless in winter when solar gains are most valuable. When designing a passive solar building the possibility of future development blocking solar access should be considered. Examples of good practice in layouts are passive solar homes at Gifford Park, Milton Keynes and passive solar homes at Willow Park, Chorley.
Good practice references and further reading:	• Duncan, I.P., and Hawkes, D., *Passive solar design in non-domestic buildings*, ETSU Report S–1110 1983. • Campbell, J., and White, K., *Passive solar design studies for non-destructive buildings*, Report 51157, ETSU, 1987. • NBA Tectonics, *A study of passive solar housing estate layout*, ETSU Report S–1126 1988. • Littlefair, P.J., *Site Layout planning for daylight and sunlight*, BRE Report 1991, Section on Passive Solar Design.

Issue D3.3.8	Earth sheltered design
Background: ▸ D4.5.2	The possibilities of building within the earth, to avoid disturbing the landscape or to restore it to its former appearance, may be an appropriate response on some sites. The great thickness of earth surrounding the building can give excellent insulation. An alternative can be to use earth over an impervious covering to provide an attractive roof covering. As well as the obvious technical difficulties of such a form of construction there are other constraints such as planning requirements and the attitude of the general public. The great advantage of integrating the building with its site and covering it is that the earth is a natural moderator of temperature. Depending on the depth of soil, it can slow the passage of heat gained or lost so that heat gained in summer will reach the building in early winter and the cooling effects on the soil in winter may not reach the building until early summer. This provides a natural 'thermal flywheel' and experience has shown that supplementary heating and cooling can be provided by passive solar means, reducing energy use.
Background references:	• Vale, B. and Vale, R., *Green Architecture – Design for a sustainable future*, Thames and Hudson, London 1991. • Pearson, D., *The Natural House Book*, Gaia Books, Conran Octopus, London, 1989. • Johnston, J. and Newton J., *Building Green – A guide to using plants on roofs, walls and pavements*, London Ecology Unit, 1993.
Good Practice: ▸ D4.5.2	*Earth-sheltered design* may be a response to the requirements of a particular site and/or brief. In designing such a scheme, consider: • whether the design of an earth-sheltered structure will enable it to blend more easily into the surroundings, or is beneficial to wildlife and the wider environment; • the possibilities of providing open space for recreation above such a structure; • building into the ground to provide a more even temperature for the internal space, as in a water supply service reservoir; • building into the ground to isolate workspaces from the external environment; • orientation of openings to benefit from solar gain and passive ventilation. Good examples of earth-sheltered design are the Sir Joseph Banks Building in Kew Gardens and the Crescent Wing extension to the University of East Anglia Sainsbury Building.
Good practice references and further reading:	• As background references, plus: • Contact the British Earth Sheltering Association, Caer Llan Berm House, Lyddant, Monmouth, NP5 4JJ about the benefits and construction of earth-sheltered structures.

D3.4 Impacts on occupants and users, visitors and neighbours

Issue D3.4.1	User and community consultation
Background: ▸ D2.3.9	People are more likely to respond favourably to their surroundings if they are able to make choices about where they live, work, shop and enjoy themselves and can have some influence in shaping the built environment. A respect for human needs and aspirations should therefore characterise all design activities. Involving users or potential users of civil engineering works and industrial buildings in the design decisions is often a normal part of the design process since those users may work for the project client such as a water services company. In other cases, the local community may be actively involved in selecting for example the least disruptive sea defences from a range of engineering solutions or in selecting the best route for a by-pass. Wherever possible people who will use, or be affected by, the scheme should be consulted during the earliest stages of the primary design process. This could take the form of positive involvement of the users in the design and construction process, at the smallest scale involving people in self-building their own homes. Community architecture may involve encouraging communities to take charge of shaping their own housing and local facilities in run down inner city areas. Involving building users is equally applicable to the refurbishment of existing buildings as to new buildings. As part of the planning process, the wider community will also be consulted about every proposed development. On larger scale constructions this process of consultation is vital to balance the many conflicting priorities.
Background references:	• Vale, B. and Vale, R., *Green Architecture – Design for a sustainable future*, Thames and Hudson, London, 1991, in particular pp 128–136.
Good Practice:	*In designing for user groups and communities, consider:* • carrying out preliminary studies to define options available technically and financially before undertaking consultations, as it is unfair to raise unrealisable expectations; • commissioning a social survey of people to establish some of their priorities and to obtain essential data about potential problems, their attitudes to the proposals for a new project and their aspirations; • the employment and training opportunities during development and on completion; • undertaking structured consultations, which may take the form of weekend workshops and evening meetings to explain key aspects of the design, funding and construction and to illustrate constraints and opportunities, in order to establish a consensus amongst participating occupants/users; • holding 'interactive' meetings, using sketch drawings and models interspersed with slide illustrations during which both the designers and occupants/users can formulate design strategies and priorities; • organising visits by users to other schemes so that they can see the results of the design process; • communicating the decision-making process in newsletters or information sheets to ensure that everyone is kept informed; • ensuring thought is given to working restrictions for the contractor and to ameliorating conditions for occupants if refurbishment work is required; • commissioning follow-up surveys after completion to assess the impact of the scheme and to provide feedback for future design work.
Good practice references and further reading:	• Wates, N., and Knevitt, C., *Community Architecture*, Penguin, London, 1987. • Tower blocks – a resource for the future?, article in *'Housing Review'*, Vol.42 No.1, January–February 1993. • *Environmental Assessment: Guide to the identification, evaluation and mitigation of environmental issues in construction schemes*, CIRIA Special Publication 96, 1993.

Issue D3.4.2	External appearance (aesthetics)
Background: ▸ D2.3.12	In the past the preponderance of vernacular buildings and civil engineering works such as arch bridges or viaducts relied upon the best use of available local materials allied to traditional construction methods. It could be argued that the resulting tradition was mere construction and that the additional ingredient of 'aesthetics', that art which makes a structure or building beautiful, was lacking. Aesthetic considerations, the delight one often sees in symmetry and order in a beautiful bridge or a vernacular building, is nonetheless part of that tradition. There is also another tradition, that of monumental structures and buildings, often entailing a profligate attitude to use of resources in pursuit of aesthetic effect. Both traditions have produced places of beauty, but it could be argued that the latter tradition is more prevalent in building than civil engineering. It should only be followed when the building celebrates some special use: a church, town hall or other building of religious or civic importance, where there is a need for it to be in keeping with adjoining buildings or when a client has a special project. A beautiful and resource-conscious public building is now needed to demonstrate that the monumental tradition can encompass good use of available natural resources and also produce art.
Background references:	No specific references identified.
Good Practice: ▸ D3.6	This is a very subjective area, since the perception of aesthetics varies from one person to another and indeed from one culture to another. However, there are simple attributes on which good architecture has always been founded, irrespective of the style, period or structure involved. These include: • scale, bulk and proportion; • relationship of the project to its surroundings; • texture of materials; • relationship of solid to void in both plan and elevation; • architecture experienced as planar surfaces; • rhythm and order; • light and colour; • where appropriate, as in concert halls, good acoustics. *In developing a structure or building in relation to its site, consider, subject to planning constraints:* • the choice of form and materials in the context of the site and surroundings; • uniting the structure or building positively with the site, planning it to retain an existing tree, rock outcrop or other site feature; • using materials made or extracted locally; • lightweight construction where disturbance to the site is minimised and materials can be re-used when the works or building has reached the end of its useful life; • creating a sense of place by emphasising any unique qualities of the site, avoiding standard solutions and acknowledging local traditions of materials and form.
Good practice references and further reading:	• Rasmussen, S.E., *Experiencing Architecture*, 1964, Chapman and Hall Ltd. • Krier, R., *Architectural Composition*, Academy Editions, London 1988. • Vale, B. and Vale, R., *Green Architecture – Design for a sustainable future*, Thames and Hudson, London, 1991, Introduction and Postscript.

Issue D3.4.3	Noise
Background:	The effect of off-site noise on a development has been considered in some detail in Stage 2 and for background information you should refer to section D2.3.7.
Background references:	See D2.3.7.
Good Practice: ▸ D4.6.3	*At conceptual design stage, recognise that:* • the selection of a site should take into account the effect of noise on the type of development proposed – for example, homes should not be sited close to airports, busy roads or noisy factories; • conversely, developments which are likely to be noisy should be sited so as to cause the minimum loss of amenity to neighbouring residential areas; • sometimes the effects of noise are not always easy to foresee, and noise can be generated or reflected in unexpected ways by new developments; • ambient noise levels on site need to be measured at the outset to determine design requirements; • barriers or grassy mounds could be used to screen a noise source such as a road or railway. *To exclude noise from a building or from rooms within a building, consider:* • in residential buildings, kitchens, bathrooms and halls can be placed on the noisy side of the structure to act as buffers; • similarly in non-residential buildings, store rooms, service areas and circulation routes should be placed to act as a buffer on the noisy side of a building; • in high rise buildings, less sensitive rooms like store rooms or kitchens can be located next to sources of noise such as lifts, lift motors and pump rooms; • when converting older buildings, particularly for residential use, the reduction of noise nuisance can be assisted by planning so that like rooms are stacked above each other or share a common wall; • external and common walls, and floors between houses and flats, should generally be of a heavy construction, although sound insulation can also be achieved in other ways, for instance, by separating party wall leaves and providing insulation quilt in the cavity as in timber frame construction; • roofs may need to be of heavyweight construction or separate leaves of lightweight construction to exclude aircraft noise; • use of gardens, open spaces, trees, shrubs and hedges should not be overlooked; • in particularly sensitive buildings such as airport terminals, other buildings near airports or busy roads and urban office buildings, recognise the importance of appropriate glazing – double or triple – in reducing interior noise; • acoustic ventilators may be needed for fresh air where windows cannot be opened. The refurbishment and extension of the Grove Road Primary School in Hounslow, West London is a good example of design to exclude aircraft noise from a building. *In civil engineering projects:* • consider as an integral part of the design the noise impact of the project on the surrounding area and seek to design out significant noise emissions across the site boundary, for example from electrical or mechanical plant; • in projects in open country or beside or near open water, recognise that noise nuisance can occur at much larger distances from the project than in more congested urban landscapes; • ensure that you and the rest of the design team are aware of any legal requirements such as those governing provision of noise barriers alongside new motorways.
Good practice references and further reading:	• Pearson, D., *The Natural House Book*, Gaia Books, Conran Octopus, London, 1989. • *Sound Control for Homes*, Joint BRE/CIRIA Publications 1993. • Silence in glass, article on Grove Road Primary School, in *Building*, 7 May 1993. • Beaman, A. L. and Jones, R. D., *Noise from construction and demolition sites – measured levels and their prediction*, Report 64, CIRIA, 1977.

Issue D3.4.4	Anticipating and minimising development impacts
Background:	A new development has the potential to impact on an existing community in a number of ways. These are discussed in D2.3.9 on Local Community. Considerations of good neighbourliness as well as the need to limit the potential for pollution should be carefully weighed at this stage.
Background references:	See D2.3.9.
Good Practice:	This is a wide area of potential concern. Consider the guidance in CIRIA SP96 and the following points for different forms of development: ***Civil Engineering:*** • specific pollution control strategies and controlling and purifying emissions, for example from sewage and waste treatment plants; • traffic generation by the development and the need to cope with increased traffic in existing roads; • noise generation; • anticipating routes used by construction traffic during the construction phase; • the potential for light pollution from floodlighting associated with some types of development, eg football stadia, athletics grounds etc. ***Buildings:*** • traffic generated; • the noise generated and the need for screening measures; • anticipating routes used by construction traffic when the building is under construction; • overshadowing and the need for privacy; • limiting fumes generated by some building types such as petrol stations; • glare caused by solar reflections from glazed facades; • weatherproofing exposed neighbouring party walls and the need for party wall notices in urban locations. ***Industrial:*** • traffic and noise generation; • air pollution from dust, smells and fumes; • water pollution from plant emissions; • anticipating routes used by construction traffic when the building is under construction; • aesthetically unsympathetic buildings and process plants.
Good practice references and further reading:	• *Environmental Assessment: Guide to the identification, evaluation and mitigation of environmental issues in construction schemes*, CIRIA Special Publication 96, 1993. • Ministry of Agriculture, Fisheries and Food, *Environmental Procedures for Inland Flood Defence Works – A guide for managers and decision makers in the NRA, Internal Drainage Boards and Local Authorities*, 1992.

D3.5 Detailed consultation with relevant bodies

Issue D3.5.1	Water quality and economy
Background: ▸ D2.3.2	The effect of a development on site hydrology and water quality has been considered in Stage 2 – refer to D2.3.2 for background information on water quality. Water is also an increasingly scarce resource with associated financial and environmental costs from the development of new sources. It is therefore important to look at ways of decreasing demand for publicly supplied water when considering development proposals.
Background references:	• See D2.3.2. • DoE and Welsh Office, *Using water wisely*, Consultation paper, July 1992.

Good Practice: ▸ D2.3.2	*In reaching design decisions related to water quality and economy on your projects:*
	• review D2.3.2 and consult the NRA if not done already;
	• identify the need for licences and/or consents:
	– a discharge consent – for contaminated 'trade effluent', such as water draining from large car parks, lorry loading bays, fuel storage and filling areas, and consent will also be required for treated sewage effluent to a watercourse or soakaway;
	– land drainage consent – for any works on the bed and banks of a river;
	– impounding licence – for any impounding of water eg creation of artificial ponds;
	– an abstraction licence – for all abstractions in excess of 20 cubic metres/day, and consent is also required for test boreholes;
	• identify means of promoting water economy on the project, for example:
	– rainwater collection for watering soft landscaped areas and amenity spaces;
	– 'grey water' collection from wastewater outlets for watering amenity spaces.
▸ D4.6.2	• review D4.6.2.
Good practice references and further reading:	• National Rivers Authority, *Pollution Prevention Guidelines Working at Demolition and Construction Sites*, NRA, July 1992. • Contact NRA in your local region.
Legal references:	• Discharges of waste water to controlled water and water contamination are regulated by the NRA under the Water Resources Act 1991. • Discharge of trade effluent into the public sewer system is controlled by water and sewage companies pursuant to the Water Industry Act 1991.

Issue D3.5.2	Floodwater provisions
Background: ▸ D2.3.2	Flooding by incursion, by rain or by deluge is a potential hazard in many locations and the potential for flooding needs to be actively considered in the design of any civil engineering or building project. Development on coastal sites and in river flood plains should have been carefully reviewed in Stage 2 – refer to D2.3.2 for background information on site hydrology. There is increasing concern that global warming will result in rising sea levels, with the result that certain localised coastal areas and river estuaries will no longer be viable locations for building or other developments.
Background references:	• See D2.3.2.
Good Practice: ▸ D2.3.2	*In relation to floodwater provisions:* • ensure careful evaluation of floodwater risks in respect of coastal sites and sites within river flood plains has been carried out, and review the results; • developments which may cause significant increase in flood risk should have had an environmental assessment carried out at Stage 2, and the results should be reviewed; • consult NRA about works likely to impede flows on the bed and banks of a river, and local byelaws which forbid obstruction or infilling of the floodplain without consent; • set building levels above known floodwater levels, or design building fabric to keep water out of areas of the building below floodwater level; • consider the creation of flood storage lakes and flood channels where appropriate; • if necessary, design basements or other below-ground structures to be watertight.
Good practice references and further reading:	• *Scope for the control of urban runoff*, CIRIA Reports 123 and 124, 1992. • *Design of flood storage reservoirs*, CIRIA/Butterworth, 1993. • Ministry of Agriculture, Fisheries and Food, *Environmental Procedures for Inland Flood Defence Works*, 1992. • Contact NRA in your local region.

Issue D3.5.3	Archaeological and historical issues
Background: ▸ D2.3.12	Archaeological and historical considerations should have been taken into account at planning and site investigation stages. At this stage it is the details of how primary design choices may influence these issues which will need consideration and the involvement of relevant bodies in the process.
Background references:	See 2.3.12.
Good Practice:	*At the earliest practicable stage:* • consult appropriately with, for example, local authority archaeologists, the ICE's panel on historic works, English Heritage, Civic Trusts, building conservationists, local historical societies and any other organisation with a legitimate interest in the site; • consider the employment of archaeological and/or building or civil engineering conservation consultants; • consider the need for and form of funding and agreements covering any archaeological investigations; • consider the impact of any such investigations on the project design and programme.
Good practice references and further reading:	• Hayward, D., Pot Luck, in *New Civil Engineer Limehouse Link Supplement*, Thomas Telford, May 1993.

D3.6 Designing appropriate landscaping

Issue D3.6	Designing appropriate landscaping
Background:	Quality in landscape design can enrich any development. However, landscape designers and architects are coming under increasing criticism for designing landscaping schemes which pay little respect to what was present on the site before the construction project began, to the site's overall landscape and ecological context and to the need to retain and reinforce those features of a place which make it distinct from another. Although there are some notable exceptions such as the planting of motorway cutting and embankment slopes, all too often the same exotic species are used in the same combination and design style to create a landscape which is similar to that used in other new developments around the country and yet sits uncomfortably in its surroundings. Generally speaking plants indigenous to the UK are better for wildlife than exotics. Elements of hard landscaping such as pavements, footpaths and car parking, the provision of lighting and signing, and possibly features such as children's play areas, balancing lakes, visual and noise barriers can also be positive features and will need expert consideration at this stage. Effective landscaping can significantly ameliorate the impact of civil engineering works and, particularly when the original site was derelict or otherwise unattractive, greatly enhance the overall benefit of the project.
Background references:	See good practice references.
Good Practice:	Landscaping can be used to screen unsightly areas, to soften the form of the development, to provide all year round visual interest, to provide physical barriers to increase security and restrict access, and to encourage wildlife.

Good Practice continued:	*In designing appropriate landscaping:*
▸ D3.3.2	• developers and landscape designers should make themselves familiar with the site and its existing landscape features; • the form and species composition of surrounding landscape should influence the landscape design for the site, particularly on its periphery; • local authority staff, particularly landscape architects, planners and ecology or countryside officers should be consulted for their advice; • consider the implications of any policies or guidance notes on the landscaping of new developments which may have been produced by the local authority or other relevant bodies; • retain any features of existing interest identified at planning and site assessment stages – for example existing hedgerows or streams should be retained and possibly used for footpath or cycle access; • recognise that good landscaping practice also includes creating *new* landscape and ecological features; • retain mature trees, particularly those protected by TPOs and use them to guide and inform design; • maintain and/or create links with existing landscape and ecological features such as woods, open space and footpaths; • consider employing an ecologist to work alongside a landscape designer on sites having wildlife interest or where habitats are to be created; • consider employing an arboriculturalist to advise on trees; • use species of appropriate scale, growth rate, shape and colour and provenance to the situation; • make the client aware that landscapes may take years to establish and mature, can change dramatically over time and will require management and maintenance; • make sure any provision of open space, particularly regarding housing development, is appropriate to the needs of end users; • give consideration to whether or not any building to be constructed is to have plants growing on its roof or walls or is to incorporate any facilities for nesting birds or roosting bats; • keep public footpaths away from road edges wherever possible; • ensure if possible that footpaths follow the most desirable and direct routes to link parts of the development, with the safety and needs of end users a paramount consideration; • include in any water retaining features of the project, such as temporary flood storage ponds, positive ecological features such as permanent wetland or reed-bed which would reduce or improve the visual impact of the works; The *Daily Mail* operates a Green Leaf Housing Award scheme which aims to: • reward, encourage and increase environmental awareness in the house building industry; • show that new housing can improve the landscape and enrich the quality of life in local communities.
Good practice references and further reading:	• Baines, C., *Landscapes for New Housing: The Builders Manual*, New Homes Marketing Board, 1990. • Johnston, J. and Newton J., *Building Green – A guide to using plants on roofs, walls and pavements*, London Ecology Unit, 1993. • Coppin, N.J. and Richards, I.G., *Use of vegetation in civil engineering*, CIRIA Book 10, 1990. • Department of Transport, *Manual of environmental assessment*, HMSO, July 1993. • *Green Capital: Planning for London's Greenspace*, Countryside Commission Publications, Manchester, 1991. A series of recommendations to guide London Boroughs in drawing up policies to protect and appropriately manage green space. • *The Daily Mail Green Leaf Awards* 1993, New Homes Marketing Board, London. A pamphlet explaining the aims and requirements of the award scheme. • *More Homes and a Better Environment*, Report by the New Homes Environmental Group. This report focuses on the problems of providing affordable housing in an attractive environmental setting. In particular it assesses the role of the planning system and what makes good design.

D3.7   Energy options

Issue D3.7.1	Energy conservation and efficiency
Background:	Atmospheric warming is caused by the absorption by some gases in the atmosphere – so-called 'greenhouse gases' – of infra-red radiation from the surface of the earth. Since the industrial revolution, with the increased burning of fossil fuels and the development of agriculture and deforestation, the levels of greenhouse gases in the atmosphere have risen. There is evidence that the concentration of greenhouse gases has increased over the last 30 years, together with an increase in global air temperatures. The main greenhouses gases are Carbon Dioxide (CO_2), Chlorofluorocarbons, which are known as CFCs, Methane (CH_4), Nitrous Oxide (N_2O) and Ozone (O_3). There is still much scientific uncertainty about the global warming resulting from the increase in greenhouse gases. However, it is widely thought that even with preventative action the concentration of greenhouses gases already in the atmosphere will lead to some warming, and there are some predictions that the average world temperature could rise between 1.5 and 4.5 degrees by the middle of the next century. This would be the greatest change in temperatures over the past 5000 years, and it could lead to dramatic changes in weather patterns and inundation of low lying areas of the world. Whatever the uncertainty of such predictions, it is generally accepted that developed countries must improve their energy efficiency and reduce the level of greenhouse gas emissions if global warming – which is one of the major environmental issues – is to be reduced. In broad terms, construction industry related energy use amounts to around 55% of national energy use and, in terms of carbon dioxide emissions, a similar percentage applies. Clearly, therefore, the construction industry can take a leading role in helping to reduce carbon dioxide emissions. Construction-related energy use falls into two categories – that used by the construction industry and its suppliers in the construction of civil engineering works or buildings, ('embodied energy') and that consumed by users during the life of the works or building. Energy used directly in the construction process is only part of the total energy embodied in a structure or building; embodied energy also includes energy used in the manufacture and transport of the building materials and in materials and fittings replaced during refurbishment. Whilst it is currently much less than the energy consumed by building users, in civil engineering it can be a very significant proportion of the total project lifetime energy use. In addition, as more energy-efficient buildings are constructed, embodied energy will increase as a proportion of total lifetime energy.
Background references:	• Shorrock, L., Henderson, G., Section 3: Energy Use, Global Warming and Climate Change, in Construction Industry Environmental Forum, *Environmental issues in construction – A review of issues and initiatives relevant to the building, construction and related industries*, CIRIA Special Publications 93 and 94, 1993. • Jones Lang Wootton, McKenna & Co, Gardiner & Theobald, *A New Balance*, 1991.
Good practice references and further reading:	• BRECSU Good Practice Guides • Organisations to contact: • BEPAC 0923 664132 • EEO 071-276 6200 • BRECSU 0923 664258 • ESTA 0453 886776 • BRE 0923 894040 • ETSU 0235 432120 • BSRIA 0344 426511 • Institute of Energy 071-580 7124 • EDAS 071-916 3891 • Watt Committee on Energy 071-379 6875

Issue D3.7.1	Energy conservation and efficiency
Good Practice:	It is important to consider at the earliest stages of design what can be done to improve the energy efficiency of civil engineering works and buildings. Addressing the energy efficiency of new works and buildings is important for the longer term but, if significant improvements are to be achieved in the short term, then it is also vital to improve the energy efficiency of existing works and buildings. The potential for energy saving in different construction industry sectors include the following measures: *Civil engineering works and construction* • careful design of energy consuming features such as pipework, pump inlets and outlets, and lighting for motorways, railway sidings and ports; • use of energy-efficient pumps, motors, valve actuators, lock motors etc and low-loss valves; • overall layout design, for example of port facilities, to minimise energy consumption by vehicles and mechanical handling equipment used to operate the port; • vertical alignment of roads, railways and airport runways to minimise energy consumption in use. *Dwellings* • good levels of loft, wall and floor insulation; • choosing energy efficient heat and light sources and appropriate controls for space and water heating; • draughtproofing and double glazing. *Industrial, public sector and commercial buildings* • insulating those buildings where excess process heat is not available; • reducing air infiltration, for instance through large doors; • choosing efficient, localised heating systems; • efficient controls; • choosing improved light sources and control systems; • restricting the use of air conditioning and choosing efficient plant; • using efficient boilers and other heating appliances and proper controls; • designing levels of insulation in excess of Building Regulations requirements where appropriate; • designing the building in terms of orientation, shape, insulation levels, shading, heating etc so that air conditioning is not required at all or only required in localised areas and the use of mechanical ventilation is minimised; • making use of daylight and more efficient light sources and controls; • taking into account cooling as well as heating requirements in designing levels of insulation; • proper energy management, both on a human scale and through building energy management systems.

▸ D3.8.1

There is a wide range of technologies which the designer could consider to improve energy efficiency, some of which are considered in further detail in the following sections of the handbook.

Energy modelling in one form or another should also be considered at the primary design stage. These can range from the simplified models such as those used in energy labelling to more complex dynamic simulation models. Information and advice on modelling is available from a number of sources including BEPAC (Building Environmental Performance Analysis Club), CIBSE, BSRIA, BRE and EDAS (Energy Design Advisory Service). EDAS offer a free 1-day consultation. Energy efficiency advice and information is also available from the Energy Efficiency Office and BRECSU (Building Research Energy Conservation Support Unit) and ETSU (Energy Technology Support Unit). Electricity companies also offer ESICHECK energy efficiency analysis.

Issue D3.7.2	Use of high levels of insulation to minimise energy use in buildings
Background:	One approach to saving energy is to insulate a building so well that the majority of the space heating demand can be met from incidental heat gains, such as the heat emitted by the occupants, the heat from appliances used within the building, and heat from lights.
	Szokolay (1980), in his Environmental Science Handbook, has suggested the following gains from activities in a typical UK dwelling:
	• Cooking 2000 kWh/year • Lighting 500 kWh/year • Refrigerator 350 kWh/year • Other appliances 650 kWh/year
	and has suggested that a house might have total gains due to occupants and their activities of 4800 kwh/year.
	'Superinsulation', as it is sometimes called, combines low U-values in the building fabric with airtight construction to reduce fabric and ventilation heat losses. If a building is made airtight, a controlled ventilation system will be required to provide air needed by the occupants and to avoid condensation, and this could incorporate a heat recovery system to permit heat from outgoing stale air to preheat incoming fresh air for ventilation.
▸ D3.7.3	In a very well insulated building the construction can be heavyweight or lightweight as preferred. In fact the large majority of well-insulated houses are built using timber frame construction as it is generally easier to incorporate high levels of insulation in the voids in timber frame walls. This contrasts with a passive building, where the need to have a good thermal mass to absorb heat gains requires heavyweight masonry construction.
Background references:	• Szokolay, S., *Environmental Science Handbook*, The Construction Press Ltd, Lancaster, 1980.
Good Practice:	A well-insulated works or building can be achieved largely by the incorporation of technical features. Achieving high insulation has little effect on the traditional concerns of the designer in terms of orientation, views, site response and building appearance. It will be appropriate where the site does not permit designing the building to take advantage of solar gains.
	In adopting such an approach, designers should actively consider:
	• high levels of insulation in the building fabric, including a U-value of 0.2 or lower for the walls and ground floor and 0.15 or lower for the roof – this is likely to mean insulation of 200 mm in walls and 300 mm in roofs; • windows to be double or triple glazed, with low emissivity glass; • achieving airtight construction, • heat recovery mechanical ventilation systems.
	The combination of very high levels of insulation in roofs and external walls and glazing with low U-values means that internal surface temperatures are very close to room air temperatures, and the room feels more comfortable as a result.
	Very high levels of insulation of the building fabric alter the balance of energy uses in the building with space heating consuming less energy than electricity for lights, appliances and the ventilation system. The designer should also therefore look to specify low energy lights, and efficient ventilating systems with good controls.
Good practice references and further reading:	• Vale, B., Vale, R: *Towards a Green Architecture, Six Practical Case Studies* – Case Study 4, 1991. • *Guide to thermal insulation and ventilation*, Publication HB812, NHBC, 1991.

Issue D3.7.3	Passive solar design of domestic buildings

Background: ▶ D3.7.2	Passive solar design exploits the potential for solar radiation to contribute to heating needs in buildings, principally dwellings. Research and demonstration work suggests that, if carefully considered and integrated into the design, solar gains can on average contribute a third of total heating needs. However, in small, well-insulated dwellings the potential for using solar energy is less because gains from occupancy can be enough to meet most of the heating requirement. The passive solar approach has found favour because it is simple, non-polluting and there are low running costs. Auxiliary heating will still be required, but the design can make a significant contribution to the reduction of this demand. Often overlooked can be the reduction in energy required for lighting because of the inherently good daylighting resulting from such a design approach.
Background references:	• See D3.3.7.
Good Practice: ▶ D3.3.6	*To maximise the benefit from passive solar heating careful attention has to be paid to the following design parameters.* • The site should be laid out to minimise overshadowing. • Buildings should be orientated to maximise the potential for solar gain, with large windows on the south and south-eastern faces and small ones elsewhere. • Dwellings should be planned with the main habitable rooms on the south side. Other types of building should be planned with working areas on the south face, but areas with high internal gains on the north side. • Good levels of insulation in the building fabric to reduce heat loads. U-values of 0.3 in walls and floors and 0.2 in roofs and below are typical. • Thermal mass within the insulated envelope to act as a heat store and keep temperatures up in the building when there is no direct sunshine. Dense masonry construction for walls and concrete slabs for ground floors, if carpet is not fitted, probably afford sufficient thermal mass. In practice, only the inner 50 mm of the building fabric has a significant effect. • False ceilings should either be omitted or designed to use the ceiling space as a plenum space if the effect of the structural soffit on thermal response is to be realised. • Protection from summer overheating, including use of overhangs or shading to exclude summer sun and positioning ventilation openings to act as a cooling thermal chimney in summer. • Careful consideration given to the type of glazing to reduce heat losses through windows. U-values of about 0.5 are technically possible with argon filled triple or quadruple glazed sealed units. • A responsive heating system with controls to minimise auxiliary heating required by turning the heating down or off as the solar gains build. • Introduction of unheated conservatory or winter garden spaces on the south face of buildings to act as a buffer, to capture and store solar radiation, and to pre-heat ventilation air into the building. Careful attention needs to be paid to shading and ventilating these spaces in summer to prevent overheating and the temptation to provide heating in them in winter should be resisted. Good examples of passive solar designed homes are houses designed for the Energy World exhibition at Milton Keynes by a number of architects.
Good practice references and further reading:	• Goulding, J.R., Owen Lewis, J., Steemers, T.C., *Energy Conscious Design: A Primer for Architects*, Batsford, 1992. • Lebens, R., *Passive Solar Heating Design*, Applied Science, London, 1980. • Mazria, E., *The Passive Solar Energy Book*, Rodale Press, 1979.

Issue D3.7.4	Renewable energy sources
Background: ▸ D3.7.3	There are a number of renewable energy sources including hydro-electric power, wind power, wave power, tidal power, geo-thermal power, bio-fuels and solar energy. In the main, renewable energy sources are mostly used for electricity generation. For the construction industry the biggest impact of these technologies are therefore likely to be in the construction of power-generating facilities. Passive solar design is one example of a renewable energy source to impact on the design of buildings. Active solar technology can have a major effect on design if adopted as the power source for remote facilities such as buoys or river flow control telemetry.
Background references:	• *Review* – The Quarterly Magazine of Renewable Energy – published by the DTI. • *Renewable Energy Publications* – distributed on behalf of the DTI by ETSU (Tel: 0235 432120).
Good Practice:	• Of the renewable energy sources, wind and solar power are particularly likely to be of interest in the design of buildings, or in the direct powering of facilities in civil engineering projects. • Siting of wind turbines close to buildings to generate electrical power is one possibility, although there are concerns about the visual impact, noise and disturbance to radio and TV signals, and turbines may be particularly cost effective if the site is in a remote area without access to normal power supplies. • Solar panels, the most common form of active solar heating, are most often installed to pre-heat water for conventional hot water supplies via a gas boiler or an immersion heater. • Wind energy can be used to pump water into header tanks, when mains pressure is insufficient as in some remote areas, and can be used to pump water through solar panels. • Electricity produced from photo-voltaic cells sited on building roofs or built into glazed areas can now be achieved at reasonable cost. • For civil engineering applications, wind turbines or photo-voltaic cells combined with re-chargeable batteries and suitable controls can prove to be a very effective way of providing power to remote installations with small power requirements or to installations where the cost or practicalities of laying conventional power supplies is more expensive.
Good practice references and further reading:	• As background references, plus: • Network for Alternative Technology and Technology Assessment, *Newsletter*, Open University, six times per year. • Centre for Alternative Technology publications, Tel: 0654 703743.

Issue D3.7.5	Combined heat and power
Background:	Combined heat and power (CHP) is the generation of electricity and useful heat in a single process. It makes use of the heat produced in generating electricity rather than discharging it to the atmosphere. It can typically operate at 70%–90% efficiency compared with 35%–50% efficiency of conventional power stations. It can make a valuable contribution to the environment by reducing carbon dioxide emissions. CHP is also flexible and can use any fuel.
Background references:	Organisations to contact: • CHPA (Combined Heat and Power Association) Tel: 071-828 4077 • ESTA: Tel: 0453 873568 • ETSU: Tel: 0235 432120

Good Practice:	• In the past, CHP schemes have not been particularly successful in the UK. Siting of commercial or residential buildings close to the power and heat source can be a major factor in their viability. • However, there is an increasing number of small-scale CHP installations, and the EEO (Energy Efficiency Office) has encouraged CHP installations where they can be shown to be cost-effective. Their use has become established in a number of building types including sports centres, hospitals and hotels, and there is a growing use in other areas such as educational and larger residential buildings. • Before embarking on a CHP project, designers should consider energy-saving measures to reduce heat and power demands which could in turn lead to a smaller and cheaper CHP installation. The local profile of power and heat demand is very important if CHP is to be viable, cost-effective and energy efficient.
Good practice references and further reading:	• EEO Good Practice Guide No.1, *Guidance Notes for the Implementation of Small Scale Packaged Combined Heat and Power.* • EEO Good Practice Guide No.3, *Introduction to Small Scale Combined Heat and Power.* • EEO Good Practice Guide No.43, *Introduction to Large Scale Combined Heat and Power.* • EEO Good Practice Guide No.60, *The Application of Combined Heat and Power in the UK Health Service.*

Issue D3.7.6	Design of civil engineering works for minimal energy use
Background:	In this context, civil engineering works can be divided into two broad groups: • facilities which are operated as a process, such as sewage or water treatment works, and • facilities which others use and, in doing so, consume energy in the process, such as roads, ports, harbours and locks on waterways. In addition, it may be possible to generate, and use on site, energy from the process for which the works are being designed, for example methane-generated power at sewage treatment works and landfill sites. At the primary design stage, engineers can significantly affect energy consumption of or on the works when completed through their design decisions.
Background references:	No specific references identified.
Good Practice: ► D3.7.4	*When making primary design decisions on civil engineering works, consider:* • the potential for using/specifying high efficiency or low-energy-consuming motors, pumps, valves, actuators etc in any process aspect of the project; • the potential for generating power on site; • if the process concerned generates energy or could do so indirectly by, for example, production of methane, so that the use of energy from outside the site is kept to a minimum; • the layout and alignment of the works to minimise the energy consumed by users of the facilities; • use of renewable energy sources.
Good practice references and further reading:	No specific references identified.

D3.8 Labelling and other environmental information

Issue D3.8.1	Use of energy labelling schemes
Background:	So far energy labelling schemes in construction have been developed only for dwellings. The procedures employed are based on versions of the Building Research Establishment Domestic Energy Model (BREDEM). Full descriptions have been published by BRE. For those designers wishing to assess a dwelling's energy efficiency and potential for savings, two audit schemes are available. The National Energy Foundation (NEF) offers a National Home Energy Rating Scheme (NHER) which rates dwellings on a scale of 0 to 10. The audit can be carried out by a licensed assessor for existing dwellings and new dwellings, using microcomputer programmes. A version of the programme is also available for stock surveys of existing dwellings for local authorities and housing Associations. MVM Starpoint also offer home energy audits which rate homes on a scale of one to five stars. The audit is again carried out by a licensed assessor. The Government recommends both schemes and, to allow comparison of the two different labelling systems, it has introduced a standard assessment procedure (SAP) which rates homes on an energy efficiency scale of 1 to 100. Both Starpoint and NEF now give SAP ratings as well as their own rating. A useful feature of both ratings is an estimate of annual carbon dioxide production from the use of the dwelling. Similar energy labelling schemes need to be developed for other building types. The inclusion of energy ratings is being considered in the next round of Building Regulation amendments, but at present there is no statutory requirement to use such audits. However, many local authorities and housing associations are now specifying energy targets to be achieved in the design of new dwellings, usually 8 or better on the NHER scale. Associations are interested in providing tenants with affordable warmth and a warm dwelling is likely to involve them in lower maintenance costs.
Background references:	• Anderson, B.R., Clarke, A.J., Baldwin, R., Millbank, N.O., *BRE Domestic Energy Model: Background philosophy and description*, R66, BRE, 1985. • Organisations to contact: 　MVM Starpoint, Bristol: 0272 250948. 　NEF, Milton Keynes: 0908 672787
Good Practice:	The National Federation of Housing Associations and BRECSU on behalf of the Energy Efficiency Office have published the results of a collaborative project examining higher standards of energy efficiency in new housing schemes. The publication *Affordable new low energy housing for Housing Associations* identified a range of design options to achieve NHER ratings of 8 and 9. ***To achieve NHER 9 the following measures are typically recommended for gas-heated dwellings:*** • 100 mm cavity wall insulation (U = 0.3); • 50 mm ground floor insulation (U = 0.35, dependant on configuration of the ground floor slab); • 200 mm roof insulation in two layers, 100 mm between ceiling joists and 100 mm across the joists (U = 0.18); • double glazed and draught sealed windows and doors; • condensing boiler and thermostatic radiator valves; • primary pipework insulation; • low energy lighting. Schemes are being designed with plans to achieve NHER 10 and a number of private housebuilders have also carried out developments achieving NHER 10. They perceive a marketing advantage in offering a more energy efficient house than their competitors. Although not a formal labelling scheme, the CIBSE Building Energy Code allows for the choice of the best energy option as a basis for design.
Good practice references and further reading:	• *Affordable new low-energy housing for Housing Associations*. Report FPRD2 from BRECSU Enquiries Bureau (0923 664258). • Low-energy housing design, article in *Architect's Journal*, 16 December 1992. • CIBSE, *Building Energy Code*, in 4 parts, CIBSE, 1975–82, Part 2.

Issue D3.8.2	Green labelling of materials and products

Background:	There is an increasing need amongst clients, designers and specifiers for information about materials and products to assist in decision-making and to help identify products which have fewer than normal harmful effects on the environment. A product environmental labelling scheme called the Blue Angel scheme has existed in Germany since 1978 though many believe it has limitations. There is a need for a similar scheme here.

Two significant developments relating to labelling of products with environmental information have recently occurred. Firstly the EC has been considering amendments to the Dangerous Substances Directive to label substances dangerous to the environment. Secondly EC Regulation 880/92 governing an eco-labelling scheme has been adopted and launched. It will be voluntary and criteria are to be set for each product category on the basis of a 'cradle-to-grave' analysis of environmental performance. Only those products which meet these criteria will be awarded a label.

The scheme is designed mainly for consumer goods rather than commercial products. However construction products will be included and paints and light bulbs have already been subject to pilot studies and work has been initiated on insulating materials. The UK Ecolabelling Board has been established to run the EC Ecolabelling Scheme in this country. Some manufacturers are starting to review the environmental impact associated with their products. |
| *Background references:* | • Construction Industry Environmental Forum, *Environmental issues in construction – A review of issues and initiatives relevant to the building, construction and related industries*, CIRIA Special Publications 93 and 94, 1993. Section 8.
• Atkinson, C.J., and Butlin, R.N., *Ecolabelling of building materials and building products*, BRE Information Paper 11/93, BRE, 1993.
• *The EC Ecolabelling Scheme*, Guidelines for Business, CBI Leaflet undated.
• Construction Industry Environmental Forum, *Life-cycle eco-analysis in building and construction*, Notes of meeting held on 25/5/93, CIRIA, 1993.
• Organisations tc contact:
  UK Ecolabelling Board: 071-820 1199
  BRE: 0923 894040. |
| **Good Practice:** | Evidence suggests that possession of an ecolabel by a product will give its manufacturer a competitive edge. Criteria for award of a label will vary from product to product both in the production and use of the product. Lack of data is likely to be a problem for some time, and the compilation of environmental information and the rationalisation of the way it is expressed would be of help to designers and specifiers.

Designers should meanwhile **seek the following information** when making primary choices about materials whilst recognising that it may not be available:

• the embodied energy in the material (or associated CO_2 emissions);
• the impacts associated with its extraction of the source material;
• the pollution associated with the manufacture of the products;
• the sustainability of the resources from which it was derived;
• the recyclability of the product/the recycled content of materials;
• emissions affecting the well-being of users during and after construction;
• problems arising from the disposal of materials after use. |
| *Good practice references and further reading:* | • See background references, plus:
• Construction Industry Environmental Forum, *Environmental issues in construction – A review of issues and initiatives relevant to the building, construction and related industries*, CIRIA Special Publications 93 and 94, 1993. Sections 1.6 and 2.3.
• Elkington, J., Knight, P., Hailes, J., *The Green Business Guide – How to take up – and profit from – the environmental challenge*, Gollancz, London 1991.
• Elkington, J., and Hailes, J., *The Green Consumer Guide*, Gollancz, 1991
• *Environmental Labelling*, DTI, London DoE/DTI (Quarterly Publication). |
| *Legal reference:* | • EC Regulation (880/92) on a Community Eco-Label Award Scheme. |

Issue D3.8.3	BREEAM

Background:	Concern for the environment has grown apace over the last decade and yet buildings, which have very large environmental effects, are often not considered. Encouraging environmentally friendlier buildings will over time substantially improve the environment. In the present climate of concern, about global warming for instance, it is of advantage to make visible that one building is environmentally better than another.
	BREEAM (Building Research Establishment Environmental Assessment Method) is currently the only established method of assessing the environmental quality of buildings. The first version of the scheme, BREEAM 1/90, was developed for assessing new offices at the design stage and was published in 1990. A revised version of the scheme for new office design, BREEAM 1/93, extending the scope of the environmental effects covered by the assessment, was published earlier in 1993. Since the publication of the scheme in 1990 about 25% of the new office stock designed since its launch has been assessed. Versions of the scheme have also been published for three other building types: BREEAM 2/91 covers superstores and supermarkets, BREEAM 3/91 new homes, and BREEAM 4/93 existing offices.
	The scheme uses independent assessors to evaluate the environmental effects of the building. At the completion of an independent assessment a certificate is issued which confirms the areas of environmental concern and criteria the building design has satisfied. Carrying out an assessment at the design stage allows improvements to be incorporated before the design is fixed. A credit on the certificate would indicate that the building will perform better environmentally than normal practice. A summary rating of fair, good, very good or excellent is included in the certificate based on a minimum number of credits achieved overall and in each of the following broad categories.
	The environmental issues concerned in the scheme are grouped into the global, neighbourhood (or local) and indoor environments. The issues covered in each version of the scheme are slightly different, but broadly follow the environmental concerns set out below:

Global issues:
- Rainforest destruction
- Resource depletion
- Global warming
- Ozone depletion

Neighbourhood or local issues:
- Limiting air pollution
- Limiting noise effects
- Legionnaires' Disease
- Local wind effects
- Limiting overshadowing
- Water economy
- Site ecological value
- Re-use of existing sites

Indoor issues:
- Hazardous materials
- Lighting
- Indoor air quality
- Ventilation
- Safety and security
- Indoor pollutants
- Legionnaires' Disease

It is intended that other issues will be included in future versions of the scheme as more evidence becomes available. Further matters being considered for inclusion are life cycle energy costs, combined heat and power (CHP), manageability, location, microclimate, volatile solvents, ventilation effectiveness, thermal comfort and sick building syndrome. On the scoring for the schemes, systematic weighting of the credit scores for their relative importance is under consideration.

A version of the scheme for industrial buildings, BREEAM 5/93, is to be launched in late 1993.

Background references:	• Prior, J.J., (Editor), BREEAM 1/93, *An environmental assessment for new offices*, BRE, 1993. • Crisp, V.H.C, Doggart, J., and Attenborough, M., BREEAM 2/91, *An environmental assessment for new superstores and supermarkets*, BRE, 1991. • Prior, J.J., Raw, G.J. and Charlesworth, J.L., BREEAM 3/91, *An environmental assessment for new homes*, BRE, 1991. • Baldwin, R., Bartlett, P., Leach, S.J. and Attenborough, M., BREEAM 4/93, *An environmental assessment for existing office buildings*, BRE, 1993.
Good practice references and further reading:	• As Background references, plus: • Environmental Assessment, article in *Architects' Journal*, 13 January 1993.

Issue D3.8.3	**BREEAM**

Good Practice:	The BREEAM scheme is voluntary and has independent assessors to evaluate the environmental effects of the building at the design stage. Many purchasers and occupiers of buildings find the notion of a 'green certificate' attractive.
	Consultations with an assessor during the primary design stage and before detailed design commences allow the design to be improved before final design decisions are fixed. Contact the Environmental Assessment Scheme at the Building Research Establishment in the first instance to arrange for an assessment to be carried out. The assessment may take 2–3 weeks and a full set of building plans, an outline building specification, details of heating and services installations and a site plan will be required.
	For each of the three building types some of the more important primary design choices to gain credits are as follows.
	BREEAM 1/93: New offices
	• For CO_2 standards it is easier to gain credits for low CO_2 production with naturally ventilated rather than air conditioned buildings.
	• For CFCs there are credits for no air conditioning or for refrigerants of low ozone depletion potential in air conditioning systems.
	• For resources and recycling the provision of a separate storage space for recyclable materials (wastepaper, card etc) in the design of the building gains a credit.
	• For local wind effects there are options for achieving an adequately sheltered microclimate.
	• There is a credit for the quality of daylighting design.
	BREEAM 2/91: New superstores and supermarkets
	• Again there are credits for measures to save CO_2 emissions by reducing refrigeration, lighting, air conditioning, heating and ventilation loads.
	• For CFC emissions credits are awarded for low ozone depletion potential refrigerants in food refrigeration systems and air conditioning.
	• There are credits for separate storage space within the store's operation for recyclable materials and for providing a multiple recycling facility for customer's use.
	• On indoor air quality there are credits for controllable natural ventilation and separately ventilated areas for smokers' use.
▸ D3.8.1	*BREEAM 3/91: New homes*
	• CO_2 production due to energy consumption is calculated using BREDEM and the National Home Energy Rating (NHER) Assessment procedure. A maximum of six credits are available for low CO_2 production.
	• On CFC emissions a credit is given for specifying insulants such as mineral fibres and/or foam insulation blown with zero ozone depletion potential gases.
	• On natural resources and recycled materials there are credits for timber and timber products from well managed regulated sources or suitable re-used timber.
	• There is a credit for providing space for separate containers for household waste recycling.
	• To discourage building on ecologically valuable sites there is a credit for building on land of low ecological value and a further credit for enhancing the ecological value of a site.
	• On lighting there are credits for quality of daylighting to kitchens and habitable rooms and for the use of low energy lighting.
	BREEAM 4/93: Existing buildings
	• Minimising CO_2 emissions forms a key section, the value of the saving being assessed using ESICHECK.
	• Building services are assessed from the point of view of no air conditioning or low ozone-depletion-potential refrigerants in air conditioning, and avoidance of the use of halons.
	• In addition to water conservation, transport is also considered as a local issue with an additional credit for buildings with good access to public transport.
	• For the first time, operation and management are taken into account and issues considered include company environmental policy, energy policy and energy efficiency targeting, planned maintenance of the building, measurement of the internal atmosphere within offices and steps taken to reduce the risk of Sick Building Syndrome.
	• A checklist of 'healthy building indicators' is included.
	The new Inland Revenue tax centre in Nottingham is the first building to achieve maximum credits in a BREEAM 1/93 assessment.

D3.9 Ventilation options in buildings

Issue D3.9.1	Air infiltration and ventilation rates

Background:	Whilst adequate ventilation is essential for health, safety and comfort of building occupants, excessive ventilation may waste energy and can cause occupant discomfort. All wasted energy contributes unnecessarily to CO_2 emissions.
	A building should be ventilated by design. Infiltration of air through cracks and gaps in the building fabric is not designed for and should be considered an energy overhead. The basis of good design is to provide an airtight building envelope and then provide controllable ventilation. In this way concerns about energy conservation and carbon dioxide emissions (arising from space heating and cooling), the use of CFCs and indoor air quality are properly addressed. The concept of 'build tight – ventilate right' is being promulgated by BRE and is reflected in the guidelines in Parts F and L of the new Draft 1993 Building Regulations.
	The average UK dwelling has twice the air leakage of the average North American dwelling and four times that of an average Swedish dwelling. Similarly a typical conventional office building in the UK has twice the air leakage of a Swedish office building. An additional problem such leaks cause is to provide a sound path for noise.
	Approximately 75% of air leakage in UK buildings could be through hidden paths in the building fabric rather than identifiable gaps and cracks in the building envelope. It is important therefore to design and construct buildings effectively to make them airtight rather than use post-construction remedial measures to rectify an already leaky building.
	With tighter buildings greater attention must be paid to providing adequate controlled ventilation. CIBSE gives guidance on fresh-air requirements relating to metabolic needs, controlling body odour (8 litres/s fresh air per person), and on tobacco smoke (from 16 to 32 litres/s per person).
Background references:	• Perera, E., Parkins, L., Build tight – ventilate right, in *Building Services*, June 1992.
Good Practice: ▸ D4.6.6	*To make new buildings tighter:* • identify a continuous air-tight envelope on drawings of the building; • choose draught-sealed windows and doors, put closers on doors and consider draught lobbies or revolving doors at entrances; • ensure good workmanship in construction to control air leakage through the envelope; • in office type buildings use the forthcoming BRE handbook 'Guide to minimising air infiltration in office buildings'; • in factory buildings ensure the main area for designing out uncontrolled air infiltration is tackled through careful detailed design of the fabric such as eaves, corner details and where cladding and masonry walls join; • ensure detailed design around components that pass through the envelope and large openings in the envelope such as loading bays; • in major refurbishments consider the use of air pressure tests and tracer-gas techniques to establish ventilation leakages and to assess possible remedial measures; • care must be taken to include necessary vapour barriers to avoid interstitial or surface condensation. For controlled ventilation measures in dwellings and non-domestic buildings guidance is given in the next sections, D3.9.2 to D3.9.5.
Good practice references and further reading:	• As background references, plus: • Evans, B., Domestic low energy: 1 New housing, in *Architects Journal*, 21/3/1990. • Jones, P., Low-energy factories: 3 Natural ventilation, in *Architects Journal*, 23/5/90. • BS1192: 1984 Part 4: *Recommendations for landscape drawings*, British Standards Institution, Milton Keynes. • CIBSE Guide, Volume A: *Design Data*, Chartered Institution of Building Services Engineers, in sections, 1982–1991. • *ASHRAE Handbook – Fundamentals*, American Society of Heating, Refrigerating and Air-Conditioning Engineers, 1989.

Issue D3.9.2	**Natural ventilation and passive stack ventilation in small buildings**
Background: ▸ D3.9.1	Ventilation rates in dwellings and related small buildings are normally based on the requirement to remove moisture from kitchens and bathrooms and to prevent condensation. The ventilation rates for the removal of moisture will generally be more than adequate for the other needs which include fresh-air to fuel burning appliances, occasional rapid ventilation to dilute pollutants and for cooling, and a reasonable level of background ventilation. All new buildings have to comply with Part F of the Building Regulations which state that 'there shall be adequate means of ventilation provided for people in the building'. The importance of controllable ventilation cannot be overstated. Draughts and infiltration of air through adventitious openings in the building fabric, cracks around window and door frames and around electrical and plumbing outlets give rise to discomfort and unnecessary heat losses. A technique for removing moisture is passive stack ventilation. This is a system which uses the natural stack effect of 100–150 mm diameter pipes as a vertical ventilation path from wet areas, such as kitchens and bathrooms, to outlets at ridge height. The warm moist air generated by activities in these rooms is extracted by the stack effect, rising up the stack to external air, and is replaced by air drawn from other rooms in the building. This can be used as an alternative to air-extraction fans in two storey and three storey buildings, and is included in the forthcoming draft amendments to Part F of the Building Regulations.
Background references:	• *Domestic draughtproofing: ventilation considerations*, BRE Digest 306, February 1986. • Stephen, R.K. and Uglow, C.E., *Passive stack ventilation in dwellings*, BRE Information Paper IP 21/89.
Good Practice:	*In the design of dwellings and smaller buildings the following aspects of ventilation need to be considered:* • design to ensure adequate natural ventilation, usually by cross-ventilation – if the design permits only single-sided ventilation, for instance in flats, consider a mechanism for temperature driven ventilation such as openings at high and low level; • provide adequate opening windows for occasional rapid ventilation; • design buildings to be well sealed and use trickle ventilators, which are important to provide fine control of background ventilation and can do much to solve condensation problems; • the use of (carefully designed) passive stack ventilation to extract warm moist air from kitchens and bathrooms as an alternative to extract fans; • mechanical ventilation will be necessary for internal kitchens, stores and WCs; • the use of unheated conservatory or winter garden spaces on the south face of a building to pre-heat ventilation air which enters the building via the conservatory – this can make a significant contribution to cutting fuel bills; • the building fabric to be well insulated and draught stripped, which in turn means careful design to ensure the correct ventilation paths are maintained.
Good practice references and further reading:	• Stephen, R.,K. and Uglow, C.,E., *Passive stack ventilation in dwellings*, BRE Information Paper IP21/89. • Vale, B., Vale, R., *Towards a Green Architecture*, Six Practical Case Studies, RIBA Publications, 1991, Case Study 2 - Ventilation.
Legal references:	• Part F of the Building Regulations 1991. • See references in D3.2.2.

Issue D3.9.3	Ventilation in large structures and buildings

Background:	A supply of fresh air is needed in larger structures such as power stations and buildings such as large business and retail units to control the level of contaminants present in the air and for odour dilution. The requirements are governed by the Workplace Health, Safety and Welfare Regulations 1992.
	The natural movement of air through the building due to wind and temperature difference can achieve this without the capital costs, fuel costs and use of electricity associated with mechanical ventilation systems. This has the effect of limiting the building depth to 10–14 metres.
	Buildings deeper than this can still be effectively ventilated by a combination of natural ventilation and opening windows at the structure perimeter with mechanical ventilation at the centre. Window design is crucial to allow ventilation deep into the space without causing discomfort to those occupants working near the windows. However achieving natural ventilation can sometimes be at the expense of noise penetration in urban areas, and these conflicts must be balanced in the design.
	In the summer, the ventilation rates needed to assist in maintaining comfortable conditions are much greater than those required in winter. These larger rates can be obtained by cross-ventilation when differences in air pressure on either face of a structure or building, resulting from wind speed and direction, encourage air movement through it.
	The effect of temperature difference also contributes to air flow through a building. Thus if the internal air temperature is higher than that outside there will be a flow of air in through the lower openings and out through higher openings. BRE have carried out preliminary studies on 10 metre deep single sided spaces where air temperature differences can assist natural ventilation. In larger buildings this stack effect can be enhanced by providing an atrium at the centre of a deep plan building. In summer, the air in the atrium will be warmed on sunny days and allowed to escape from the top of the atrium, increasing air movement through the surrounding spaces.
	Similar ventilation can be achieved by using access or other voids as a means of encouraging air movement up through a building by stack effect. However consideration needs to be given to fire safety if this is proposed.

Background references:	• BS5925: 1991 *Ventilation principles and designing for natural ventilation*, BSI. • Potter, I. N., *Ventilation effectiveness in mechanical ventilation systems*, TN01/88, BSRIA, 1988. • Jones, P., *Ventilation Heat Loss in Factories and Warehouses*, BSRIA TN 7/92.

Good Practice:	There is a user preference generally for daylight, views and openable windows. The preferred option currently put forward by BRE is also to design new schemes so that mechanical cooling is not required.
	In designing large structures and buildings:
	• ensure that the design (and likely future changes) do not introduce obstructions to natural vent paths; • evolve the ventilation strategy with the design of the building; • consider limiting building depths to allow cross-ventilation where the site and brief permit; • consider whether providing offices on both sides of a central corridor will limit the possibilities of cross-ventilation; • introduce courtyards, light wells or atria at the centre of deep planned buildings to encourage or permit cross-ventilation; • consider use of generous openable windows to provide air and light with openable windows at high level for cross-ventilation deep into the space, and opening sashes at lower level for use by occupants sitting close to the windows; • tall floor-to-ceiling heights with exposed soffits can assist cross-ventilation and provide thermal capacity to help limit overheating in summer; • auxiliary mechanical ventilation may still be necessary to supplement natural ventilation in deep spaces and to provide background ventilation in naturally ventilated spaces;

Good Practice continued:	• mechanical ventilation will be necessary for internal kitchens, stores and WCs; • consider provision of heat reclaim and night time forced air-cooling, which can be essential in summer for non-air-conditioned buildings.
Good practice references and further reading:	• Energy Efficiency Office Best Practice Programme. • BRECSU, Good Practice Case Study 13: NFU Mutual and Avon Group HQ, 1991. • BRECSU, Good Practice Case Study 20: Refuge House, Wilmslow, 1991. • *Principles of natural ventilation*, BRE Digest 210, February 1978.
Legal references:	**EC** • Directive on the minimum safety and health requirements for the workplace (89/655/EEC). **UK** • Workplace (Health, Safety and Welfare) Regulations 1992 (SI 1992 No 3004).

Issue D3.9.4	Mechanical ventilation and heat recovery
Background:	The heat lost by ventilation can be the largest element in the heat loss from a building. The basic air-tightness of the structure should therefore be of a high standard and the use of mechanical ventilation can ensure a controlled air flow in the building to remove stale air and moisture and prevent the build-up of contaminants. One method of further reducing ventilation heat loss is to combine mechanical ventilation with heat recovery. Typically warm moist air may be extracted from wet areas, such as kitchens, bathrooms and shower rooms and passed through a heat exchanger where its heat is used to warm incoming fresh air. The net effect can be to reduce the heat loss equivalent of the air change rate to much less than its normal heat loss. In a large building heat reclaimed from exhaust air can also be used to pre-heat incoming fresh air.
Background references:	See good practice references.
Good Practice:	*In small buildings the following considerations apply:* • select method of construction to achieve a high standard of air-tightness, which can be easier with factory constructed timber frame than traditional masonry construction; • modelling of ventilation heat losses should be carried out to ensure that the provision of a mechanical ventilation and heat recovery system will cut heating costs by significantly more than the fuel costs used to run the system; • for kitchens, bathrooms and toilets, mechanical ventilation will be required under Part F of the Building Regulations; • location of the plant and routing of ducts should be considered at the outset. In larger civil engineering structures and buildings in the commercial and institutional sector, plate heat exchangers and cooling are likely to be the key ventilation and energy issues. Nevertheless heat recovery from heat wheels, which extract heat from exhaust air to preheat incoming air, can help to reduce fuel use.
Good practice references and further reading:	• Jones, P., *Ventilation Heat Loss in Factories and Warehouses*, Technical Note 7/92, BSRIA, 1992. • *Selection of air-to-air heat recovery systems*, TN 11/86, BSRIA, 1986. • *The Commissioning of Air Systems in Buildings*, BSRIA, AG3/91. • CIBSE Guide, Volume A: *Design Data*, Chartered Institution of Building Services Engineers, in sections, 1982 - 1991, Section A4 Air infiltration and natural ventilation. • Hall, K and Warm, P: '*Greener Building*' Section 5, Page 20 (AECB) 1992, for useful information on suppliers of mechanical ventilation systems with heat recovery. • Vale, B., Vale, R., *Towards a Green Architecture*, Six Practical Case Studies, RIBA Publications, 1991, Case Studies 4, 5 and 6, Sections on Ventilation.

Issue D3.9.5	Air conditioning

Background:	Deep plan buildings can make air conditioning a necessity. Frequently such buildings have been a response to the need to maximise the development potential of expensive and noisy city centre sites. Air conditioning systems have the following associated environmental problems: • many refrigeration plants in the past have utilised CFCs (chlorofluorocarbons) as refrigerants, and reduction of CFC leakage into the atmosphere is vital to prevent further damage to the ozone layer; • electricity use by fan-assisted air movement systems and chillers contributes to excessive energy consumption; • paradoxically, whilst ventilation rates in an air-conditioned building are generally higher than in a naturally ventilated building, air-conditioned buildings sometimes give rise to more complaints of occupant discomfort because they offer little individual control; • the use of wet cooling towers in association with air-conditioning has resulted in some instances of Legionnaires' Disease, where regular maintenance of the wet cooling tower and proper application of water treatment systems has been neglected. It is now considered desirable that, in the design of new buildings, the use of mechanical cooling should be kept to a minimum primarily because the energy consumption can be two to three times that of a non-air conditioned building (although good air-conditioned buildings can have energy consumptions comparable to conventionally-heated buildings). In the 1993 draft amendments to Part L of the Building Regulations, the need to make a case for the use of air-conditioning for new buildings has appeared for the first time, though only for buildings larger than 500m^2.
Background references:	• Curwell, S.R., March, C.G. and Venables, R.K., (Eds), *Buildings and Health, The Rosehaugh Guide*, RIBA Publications, 1990, indexed references to air-conditioning. • *CFCs in Buildings*: BRE Digest 358 - New Edition Oct 1992 and Addenda Feb 1993. • CIBSE Guidance Note on CFCs, HCFCs and HFCs. • *Refrigeration and the environment – typical applications for air conditioning*, BSRIA TN15/92, 1992.
Good Practice:	In new buildings design factors to reduce dependence on air conditioning and mechanical cooling are summarised in Buildings and Health (pp 443 and 444): • narrow plan forms or deep plans employing atria, courtyards or lightwells that permit daylight and natural ventilation or a combination of natural and simple mechanical ventilation; • provision of adequate levels of daylight, but with shading to prevent sun penetration to the interior in summer; • appropriate amount of external glazing to total area of external walls; • high thermal mass; • ventilation heat recovery. Air conditioning may still be essential for some buildings, where unacceptable external air and noise pollution levels prevent provision of opening windows, or in some parts of a building with excessively high heat gains such as computer rooms. For system design: • avoid CFC-refrigerant systems and, where possible, HCFC-refrigeration systems; • allow generous access space for ease of maintenance; • provide for effective maintenance of cooling towers and evaporative condensers as specified in CIBSE Technical Memorandum 13; • consider use of absorption chillers, especially where there is a source of waste heat from combined heat and power.
Good practice references and further reading:	• As background references, plus: • Halliday, S.P., *Environmental Code of Practice for Buildings and Their Services*, BSRIA, 1994. • Butler, D. J. G., *Guidance on the Phase-out of CFCs for owners and operators of Air Conditioning Systems*, PD25/93, Building Research Establishment, 1993. • Chartered Institution of Building Services Engineers, *Minimising Risk of Legionnaire's Disease*, Technical Memorandum 13, CIBSE, 1992.

Issue D3.9.6	Legionnaires' Disease
Background: ▸ D4.6.8	Background information about Legionnaires' Disease is given in D4.6.8. The lack of wet cooling tower maintenance and proper application of water treatment systems can give rise to outbreaks of Legionnaires' Disease in the area immediately surrounding the building. Wet cooling towers can be used provided that they are well designed for effective maintenance for the cooling tower and evaporative condensers as required by CIBSE Technical Memorandum 13. Alternatively, it is possible to avoid the problem by using dry cooling tower systems which, although they are less efficient and have higher energy costs with greater CO_2 emissions, do have lower maintenance costs.
Background references:	• Curwell, S.R., March, C.G. and Venables, R.K., (Eds), *Buildings and Health, The Rosehaugh Guide to the Design, Construction, Use and Management of Buildings*, RIBA Publications, 1990, Chapter C3. • BSRIA Technical Applications Manual TA1/1990, Parts 1–3, BSRIA, 1990. • Brundritt, G.W., *Legionella and building services*, Butterworth-Heinemann, 1992.
Good Practice: ▸ D3.9.5 ▸ D4.6.8	To mitigate potential problems, dependence on air conditioning and mechanical cooling should be reduced by following Good Practice Guidance in D3.9.5. If air conditioning or mechanical cooling is necessary: • provide for effective maintenance of cooling towers and evaporative condensers as specified in CIBSE Technical Memorandum 13; • consider tower location in relation to fresh air intakes and flue outlets, with housing to offer full and easy access, and with the pond of condensed water shielded from direct sunlight; • consider avoiding the problem by applying dry cooling tower systems whilst recognising that they will use more energy. For good practice guidance on minimising the risk of Legionnaires' Disease in the internal building environment see D4.6.8.
Good practice references and further reading:	• Halliday, S.P., *Environmental Code of Practice for Buildings and Their Services*, BSRIA, 1994. • Chartered Institution of Building Services Engineers, *Minimising Risk of Legionnaires' Disease*, Technical Memorandum 13, CIBSE 1992.

D3.10 Daylight and need for artificial light

Issue D3.10	Daylight and need for artificial light
Background: ▸ D4.6.5	Good lighting is essential for visual comfort in the home and workplace. There is also a general preference for daylight as opposed to artificial light. Designing to maximise the use of daylight is therefore desirable both in terms of occupant preference and also to reduce electricity use for lighting. Although the eye is highly adaptable, it tends to adapt to the average brightness it sees and if areas are excessively bright or dark in relation to the average brightness then some discomfort will occur. Excessive brightness causes three types of glare: • disability glare – caused by direct intense light; • discomfort glare – caused by long term glare from windows and light fittings which are too bright compared with the task being viewed; • reflected glare – seen in polished surfaces or in display screen equipment. Thus the balance between daylight and artificial light is important. More detail about this is given in D4.6.5 – Lighting. The psychological importance of windows for both views out and awareness of weather conditions, as well as to provide natural light, should not be overlooked.
Background references:	• Curwell, S.R., March, C.G. and Venables, R.K., (Eds), *Buildings and Health, The Rosehaugh Guide to the Design, Construction, Use and Management of Buildings*, RIBA Publications, 1990. – Appendix 2.2 on Lighting Comfort. • *CIBSE Code for Interior Lighting*, Chartered Institution of Building Services Engineers, 1984.
Good Practice: ▸ D3.3.6	The criteria for achieving good daylighting indoors are set out in Appendix C of a BRE Report *Site layout planning for daylight and sunlight* and include: • design to achieve an average daylight factor* of 5% or more if there is no supplementary electric lighting, or 2% or more if supplementary electric lighting is provided; • in dwellings a minimum average daylight factor of 2% in kitchens, 1.5% in living rooms and 1% in bedrooms is required; • when a room is lit by windows in one wall only, the depth of the room from the window wall is limited by the width of the room, window head height and surface reflectances in the rear part of the room if a predominantly daylit appearance is required, and in practice, useful daylighting may not reach more than about 5 metres from a window; • there should be no significant room areas (20% or more) or any fixed work surfaces or tables, from which the sky cannot be seen from table top height; • all of the above criteria need to be satisfied if the whole of a room is to appear adequately daylit. In addition, selection of window sizes and position is an important aspect of avoiding glare and discomfort to occupants, particularly in non-domestic buildings. (* The average daylight factor, under standard overcast sky conditions, is the ratio of average illuminance on a horizontal reference plane within a room, expressed as a percentage of the unobstructed external illuminance.)
Good practice references and further reading:	• BS8206 1992: *Code of Practice for Daylighting*, BSI, Milton Keynes. • Chartered Institution of Building Services Engineers. *Applications Manual: Window Design* - London, CIBSE, 1987. • Littlefair, P.J., *Site Layout Planning for Daylight and Sunlight*, BRE Report 1991, Appendix C: Interior Daylighting Recommendations and Appendix D: Plotting the no-sky line.
Legal references:	• Workplace (Health, Safety and Welfare) Regulations 1992 (SI 1992 No. 3004).

D3.11 Occupant comfort

Issue D3.11	Occupant comfort
Background: ▸ D4.6.9 ▸ D3.4.3	Occupant comfort is a response to the thermal, visual and acoustic environment in a building. For the thermal environment, designers may only be able to achieve 80–95% occupant satisfaction in terms of the combination of air temperature, mean radiant temperature, air movement and prevailing relative humidity within the space. Lighting requirements for a specific task are variable from person to person, age being a prime factor. Of significant importance is occupants' need to have natural light in their workspace and reference to the outside environment through a window. Peoples' susceptibility to noise varies depending upon what they are doing. At times people need to come together to communicate, at other times they need quiet for privacy or concentration on the task in hand. Intrusive outside noise and intrusive noise from environmental systems within the building can be prevented with good design and suitable attenuation measures.
Background references:	• Nicholl, A., Bradley, A., and Home, R., Appendix 2. Guidance on Thermal Comfort, Lighting and Noise, in *Buildings and Health, The Rosehaugh Guide*, RIBA Publications, London 1990. • Austin, B., Lush, D., Sections 6.3.5, 6.3.6 and 6.3.7 of *Environmental issues in construction – A review of issues and initiatives relevant to the building, construction and related industries*, CIRIA SPs 93 and 94, 1993.
Good Practice: ▸ D3.10 ▸ D4.5.4 ▸ D3.4.3	*For thermal comfort* during sedentary work involving little physical effort the designer should use the following criteria: • indoor air temperature in winter of 19–20°C, and in summer between 20°–24°C; • surface temperatures of walls etc within a space should not differ by more than 2°–3°C from the air temperature; • relative humidity of room air should not fall below 30% in winter and in summer should vary naturally between 40% and 60%; • air velocity at head and knee levels should not exceed 0.2 metres per second. *For lighting comfort:* • design to achieve good daylighting; • design to avoid glare; • consider mock-ups or other visualisation aids to investigate requirements of visual comfort such as colour, texture and brightness pattern which are not subject to an analytical approach. *For aural comfort:* • first ensure that unacceptable structural transmission does not occur; • design out intrusive outside noise sources; • locate ventilation plant rooms away from spaces in which people live or work; • select ventilation fan blades to avoid annoying frequencies and use resilient mounts; • attenuate noise travelling in ducts by changing shape and cross-sectional area of ducts and lining with sound absorbent materials; • be aware that complaints have been registered in some air-conditioned buildings that spaces are too quiet.
Good practice references and further reading:	• As background references, plus: • Curwell, S.R., March, C.G. and Venables, R.K., (Eds), *Buildings and Health, The Rosehaugh Guide to the Design, Construction, Use and Management of Buildings*, RIBA Publications, 1990. Appendix 2, p482. • CIBSE: 081-675 5211. • BSRIA: 0344 426511.
Legal references:	• Fuel and Electricity (Heating) (Control) Order 1974 and Amendment Order 1980. • Workplace (Health, Safety and Welfare) Regulations 1992 (SI 1992 No. 3004).

D3.12 Criteria for primary material selection

Issue D3.12.1	Material for the primary structure

Background:	In an ideal situation, designers of any construction project would be able to include in their comparisons of alternative designs the environmental impact of the materials chosen, in particular the material for the primary structure. Thus, in this ideal world, an environmental choice could be made between timber, steel or concrete for piles and decking for a jetty, between timber or bricks for houses or between steel or concrete frames for office buildings, although it is acknowledged that environmental factors are not the only issues affecting such decisions.
	We are not in that ideal world. CIRIA, under the auspices of the Construction Industry Environmental Forum, has started a project to clarify the present position on the environmental impact of a range of construction materials; even after it reports, many comparisons will continue to be fraught with uncertainty, since the task is very difficult.
Background references:	• *Measurement of energy consumption and comparison with targets for existing buildings*, CIBSE, 1982. • Construction Industry Environmental Forum, *Environmental issues in construction – A review of issues and initiatives relevant to the building, construction and related industries*, CIRIA Special Publications 93 and 94, 1993. • Construction Industry Environmental Forum, *Green buildings: the designer's perspective*, Notes of a meeting held on 8/12/92, CIRIA, 1993. • Jones Lang Wootton, McKenna & Co and Gardiner & Theobald, *A New Balance – A guide for property owners and developers*, 1991.
Good Practice:	Loads, strength, spans, cost, method of working and a range of other factors will dictate choice of primary material. Comparing such materials on environmental grounds is difficult and requires careful interpretation.
	Two factors that can, with caution, be used for comparison of different primary structural materials are embodied energy and CO_2 emissions. The CIBSE Building Energy Code quotes the following: • Wood: 0.1 kWh/kg or 0.36 GJ/t • Concrete: 0.2 kWh/kg or 0.72 GJ/t • Steel: 10 kWh/kg or 36 GJ/t.
	Other estimates of embodied energy for wood are very much higher. For example, a study for BRE has estimated that it could be as high as 10 GJ/t because of the high energy needed to kiln dry, preserve and transport most of the UK's sawn softwood from overseas forests. For concrete in a building frame, BRE estimates embodied energy at 1.3 GJ/t and steel at 8 to 30 GJ/t depending on the proportion of recycled material used in its manufacture. In terms of CO_2 emissions timber has the advantage that the CO_2 absorbed during tree growth can be used to offset the CO_2 emissions during processing. At a global level, timber from a sustainable source will cause less emission than either concrete frame or steel frame. There is much controversy about which of these choices causes a lower emission but, in practice, a well-designed office building is likely to have an embodied energy of 4 to 5 GJ/m^2 gross floor area and about 500kg CO_2 emission per m^2. BRE/DLE have produced an estimating programme for office designs.
Good practice references and further reading:	• Burton, T., Stoker, C., *Capital Cost/Environment Saving – Striking a Balance*, Gardiner and Theobald. • Howard, N.P., Energy In balance: Energy Embodied in UK Housing, in *Building Services*, April 1991. • Butler, D., Howard, N.P., Life Cycle CO_2 Emissions: From the Cradle to the Grave, in *Building Services*, 1992. • CIBSE, *Building Energy Costs*, in 4 parts, 1975–82.

Issue D3.12.2	Brick, block and masonry construction

Background:	Bricks are a traditional walling material. They are durable, resilient and easy to handle. Western Europe has made more use of brick than any other area in the world. Brick and masonry construction has generally been adopted in the UK since the mid-17th century, largely to avoid the risk of fire.

Clays used in brick making are extracted using conventional open-cast mining techniques. Kiln firing of bricks uses considerable amounts of energy and there may be air pollutants resulting from the firing process. However, some clays used are exothermic and require less-than-average energy input.

The use of concrete blocks has largely superseded brickwork in the construction of internal walls, including the internal leaf of external walls. The manufacture of dense concrete blocks is generally accepted as using far less energy than 'aerated' concrete blocks which give low U-values in external walls. The dense blocks can be produced close to the point of aggregate extraction and are generally easily obtainable locally, reducing transport fuel use and costs. Which of the two is better in energy consumption terms will depend on site location, local availability of the blocks and the likely life and energy use of the final building or works.

In addition, insulation can now be so effective that there is an increased possibility of the outer leaf bricks freezing and spalling in winter.

Local stone is an aesthetically pleasing choice of material in some areas of the country. However there may be health problems associated with quarrying and working the material. Also quarries are regarded by many people as disfiguring the landscape and there can be high energy costs involved in quarrying, handling and transporting stone. Apart from expense, stone may not weather well when removed from its locality, and water and frost may damage it unless protected by overhanging roofs, copings and drips. Limestone, in particular, can suffer damage from acid rain. |
| *Background references:* | • See good practice references, plus:
• Pearson, D., *The Natural House Book*, Gaia Books, Conran Octopus, London, 1989. |
| **Good Practice:**

▸ D3.12.5

▸ D3.7.3

▸ D3.7.3 | Aesthetic value and durability are likely to be prime considerations in the selection of external walling materials. Local planning requirements will also heavily influence the selection.

When choosing bricks for walling, because of the high energy consumption required by brick production, the designer should ensure that:

• any brick chosen is totally suitable for the particular location;
• it is durable and has a suitable degree of frost resistance, particularly if in the outer leaf of well-insulated buildings;
• the clay should preferably be extracted and the brick manufactured in the same region as the project in order to reduce transport costs and associated emissions;
• consideration is given to the re-use of second-hand bricks where appropriate.

The use of stone is most likely to be appropriate in renovation or repair work, where emphasis should be on using salvaged stone and matching original work. Because stone is generally easy to cut and shape it can be used for mouldings, window and door surrounds and lintels, and it is more appropriate to use it in this limited way than as a general facing material in new work.

For internal walling, make careful assessments of the most appropriate blocks to use, seeking appropriate energy and other environmental information from manufacturers. Consider also whether using separate insulants may provide a better overall solution.

One further point worth noting is that certain design approaches require a large thermal mass in the building fabric and masonry construction is therefore appropriate. |
| *Good practice references and further reading:* | • Fox, A. and Murrell, R., *Green Design: A guide to the environmental impact of building materials*, Architecture Design and Technology Press, 1989.
• Hall, K., and Warm, P., *Greener Building Products and Services Directory*, AECB, Second Edition, 1993 -Section 4, Building Fabric, Walls. |

Issue D3.12.3	Cladding materials
Background:	The use of cladding materials and light industrial wall constructions for the external wall of framed buildings has been an option for building designers since the last century. In a building with a framed construction the choice of materials for the external skin can be considered in relation to weather tightness, thermal performance, appearance, ease of maintenance and durability, without the need to consider structural load-bearing capabilities of the wall. Generally, light cladding constructions use materials with a high embodied energy content – aluminium, steel, or glass – but may use much less energy in transport because of their lower self-weight. Quite often composite insulated panels with steel or aluminium facings are chosen. However, the embodied energy of some light industrial wall constructions can be less than masonry external walls. Against this has to be balanced the shorter life and higher maintenance requirements of lightweight cladding systems.
Background references:	• *Vale, B., Vale, R.,* Towards a Green Architecture, *Six Practical Case Studies,* RIBA Publications, 1991, Case Study 3. • Data to be published by BRE and Davis Langdon and Everest on embodied energy in building materials and components.
Good Practice:	*As with the choices for masonry construction, to justify the high embodied energy content of most cladding systems or materials, the designer should ensure that:* • the cladding chosen is appropriate for its location and exposure; • insulation values are good, with U-values of $0.25W/m^2K$ or lower, although the use of CFC-blown insulation in composite panels is forbidden; • the durability of the cladding and the maintenance requirements are fully investigated, with a design life of at least 30 years being desirable; • materials are considered against environmental criteria (such as embodied energy and recyclability) if possible and a system selected which minimises overall impact; • proper consideration is given to jointing between panels and the provision of thermal breaks between the outer and inner faces of metal profiles; • the desired acoustic and weather tightness properties are not compromised by the above considerations. Aesthetic and planning considerations will also, of course, be important in the selection of wall claddings.
Good practice references and further reading:	• See background references, plus: • Fox, A. and Murrell, R., *Green Design: A guide to the environmental impact of building materials*, Architecture Design and Technology Press, 1989.

Issue D3.12.4	Roofing materials

Background:	There is a large variety of roofing materials from which to select a roof finish. Whilst the selection of a roofing material will result from aesthetic, cost and planning considerations, it is worth considering a number of alternatives and the environmental considerations in their use. Selection on the basis of a full environmental analysis will have to await the outcome of the CIRIA Project mentioned in D3.12.1.
	Natural slate or stone – Cost can be prohibitive, but these materials can be very acceptable from an environmental viewpoint, and more so if won near the development.
	Artificial slate – Artificial concrete slates, although visually inferior to natural slates, are manufactured in the same way as other concrete tiles and their use has no particular environmental problems.
	Clay tiles and accessories – Clay tiles share the characteristics of brick and, although they involve medium energy use in manufacture, are durable and available in wide range of pleasing colours and shapes.
	Concrete tiles and accessories – Usually much cheaper than slates or clay tiles, they again involve medium energy use in manufacture, but are durable and available in a wide range of colours and shapes. There are no particular environmental problems associated with their manufacture.
	Metals – Commonly used metals are stainless steel, aluminium, copper, lead and zinc. Extracting and processing of all metals involves high energy use, particularly so in the case of aluminium. It is worth noting however that metals can be recycled and most easily so in the case of aluminium.
	Roofing membranes – Waterproofing membranes are often manufactured from synthetic materials including pvc sheeting, glass reinforced polythene, synthetic rubber and bituminous material reinforced with polyester. The manufacturing processes generally involve considerable energy use. Naturally-occurring compounds such as asphalt or bitumen are lower-energy materials of considerable durability and the potential exists for recycling.
Background references:	• Fox, A. and Murrell, R., *Green Design: A guide to the environmental impact of building materials*, Architecture Design and Technology Press, 1989. • Hall, K and Warm, P., *Greener Building Products and Services Directory*, AECB, 1993. • Consult major manufacturers.
Good Practice: ▸ D3.3.8 ▸ D4.5.2	If there are no overriding aesthetic or planning considerations, then the choice of natural slate or stone, clay tiles or concrete tiles is, at present, very difficult to make on environmental grounds. A turfed roof or planted roof may be particularly appropriate on an overlooked site with the advantage of providing greenery and insulation. The designer may also wish to consider shingles or thatch on smaller buildings and where there is a reduced risk of fire damage. Metal roofing materials have a high embodied energy, and consideration should be given to durability and the potential for recycling if they are used. The use of lead roofing or lead valleys must be avoided if rainwater is to be collected from the roof for watering vegetable gardens and greenhouses, but has many advantages for other uses. As manufacture of waterproofing membranes often involves considerable energy use, designers should ensure the material is totally suitable for its particular location. Membranes containing naturally-occurring compounds such as asphalt or bitumen should also be considered. Upside-down and planted roofs can extend the life of the membrane as it is protected from weathering and thermal movement by layers of insulation and paving or earth. However such roofs must be carefully detailed and specified.
Good practice references and further reading:	• As background references, plus: • Johnston, J. and Newton J., *Building Green – A guide to using plants on roofs, walls and pavements*, London Ecology Unit, 1993, Chapter 8, Green Roofs.

Issue D3.12.5	Timber

Background:	In principle, timber is one of the most ecologically sound construction materials. Wood is a natural moderator of the indoor climate, it breathes and assists ventilation, it stabilises humidity and filters and purifies the air, it is warm to the touch and absorbs sound, it is easy to work and has a high strength to weight ratio.
	The abundance of forests in the past together with its natural properties resulted in timber technology developing rapidly in Europe and the UK. It has been and remains extensively used in civil engineering, for example in harbours and jetties. It was used for floor and roof spans in the building of churches, monasteries and palaces, as well as developing into a framing material for timber framed houses during the Middle Ages. However, continuous tree-felling without adequate re-forestation has reduced forest extent very considerably. In the developed world, forests are being damaged by pollution.
	The tropical rain forests in Africa, Asia, Malaysia and South America have absorbed huge amounts of CO_2 from the atmosphere and it is generally accepted that tropical de-forestation is adding significantly to CO_2 levels in the atmosphere, contributing to global warming. The problem now is how to distinguish good and bad, i.e. sustainable and unsustainable timber production.
Background references:	• Callister, D. J., *Illegal Tropical Timber Trade: Asia-Pacific – Species in Danger*, A Traffic Network Report, Traffic Network and WWF, 1992. • Knight, A., *B&Q's Timber Policy Towards 1995: A Review of Progress*, B&Q, 1992. • BRE, *Wood-based panel products – their contribution to the conservation of forest resources*, BRE Digest 375, September 1992.

Good Practice:
 	
▸ D4.7.5
	In considering timber as a primary construction material:
	• recognise that timber has a number of advantages as a construction material:
	  – much pre-assembly is possible so that quick erection of a shell is possible;
	  – dry construction with no long drying-out periods;
	  – high durability if well-maintained;
	  – unless it has been transported very long distances, it will have lower embodied energy than alternative materials;
	  – it is generally easier to obtain high levels of insulation;
	• recognise that, despite the concerns about timber framed buildings a few years ago it has been demonstrated that only simple measures are necessary to stop moisture getting into the frame, including the use of a vapour check on the inside face;
	• recognise that externally, either a breather membrane or sheathing in association with lightweight cladding, or sheathing with a cavity and a masonry outer leaf, will allow moisture vapour to escape.
	• consider other design uses such as unsawn poles, log buildings for small scale leisure and domestic buildings, or glue-lam for larger structures (though be aware of the environmental concerns about the adhesives involved);
	• in civil engineering applications such as fendering, actively seek to use alternatives to tropical hardwoods such as temperate hardwoods, composites, or synthetic materials;
	• recognise that, although timber is a renewable, natural material, some timber resources are being depleted at an alarming rate;
	• therefore only specify tropical timber for those purposes for which there are currently no other timbers or wood products with comparable technical properties and performance, or other materials that are not a greater environmental hazard;
	• consider specifying timber products given in the AECB directory wherever practical;
	• in refurbishment, actively consider re-use of existing timber or use of recycled timber from demolished, similar structures.

Good practice references and further reading:	• CIEF, *Purchasing and specifying timber*, Notes of meeting held on 23/02/93, CIRIA. • CIEF, *The use of timber in construction*, Notes of a meeting held on 23/9/93, CIRIA, 1993. • *Timber: Types and sources*, Publication L296, Friends of the Earth, 1993. • Fox, A. and Murrell, R., *Green Design: A guide to the environmental impact of building materials*, Architecture Design and Technology Press, 1989. • Pearson, D., *The Natural House Book*, Gaia Books, Conran Octopus, London, 1989. • Hall, K & Warm, P., *Greener Building Products & Services Directory*, AECB, 1993.

Issue D3.12.6	Services and materials

Background:	Building services are the medium by which energy and water waste streams enter and leave buildings and are crucial to the environmental impact of the built environment. However, much wastage could be avoided by a multi-disciplinary approach which seeks to integrate the architecture, structure and services strategy.
	The building services industry is also responsible for significant use of materials and release of environmentally unfriendly chemicals. The industry is actively addressing these issues having prepared a draft code of practice which addresses the life cycle of buildings and which has been piloted in a range of commercial, public and domestic construction management and demolition projects during 1993. The code is due to be published in its final form in early 1994.
	It is important to review the use of materials in building services systems to maximise use of environmentally friendly, recyclable and local materials and components. Efficient use, long life and re-use are all methods of reducing chemical waste and maximising the value of embodied energy.
	In electrical installations, copper can be easily recycled. However most switchgear and accessories use plastics, which are difficult to recycle. For mechanical services installations, metal and plastic pipes, ducts, accessories and components can be repaired, maintained and recycled if sufficient consideration is given to these requirements at the design stage.
▸ D3.13.4	The importance of designing services with easy access for maintenance and repair cannot be overstated.
Background references:	• Halliday, S.P., *Building Services and Environmental issues – The Background*, BSRIA Interim Report, April 1992.
Good Practice:	*In selection of materials and products:* • all those involved in design of services and their integration with the rest of the works or building should review and, if appropriate, work in accordance with the BSRIA code; • designers should seek advice on the environmental impact of materials and chemicals produced in manufacturing; • consider the life cycle status of products, including embodied energy, energy efficiency in use, pollution hazards during extraction and manufacture, recyclability of the product and minimising wastage; • make provision for good access for maintenance, repair and replacement of all service installation.
Good practice references and further reading:	• Halliday, S.P., *Environmental Code of Practice for Buildings and Their Services*, BSRIA, 1994. • CIBSE Guide, Volume A: *Design Data*, Chartered Institution of Building Services Engineers, in sections, 1982 – 1991. • CIBSE Guide, Volume B,: *Installation and Equipment Data*, Chartered Institution of Building Services Engineers, 1986.

Issue D3.12.7	Insulants

Background:	Insulants in the building fabric are an essential means of conserving energy and the choice of insulants during the primary design stage can be an important consideration.
	The thickness of insulation and its location in the building fabric should be considered at an early stage as it will affect the subsequent design detailing. It is also necessary to understand the characteristics and limitations of insulants before making design decisions which will influence the final choice of insulation type. For instance, when considering flat roof construction, if an inverted warm deck roof is to be used, then extruded polystyrene (XPS) foam is the insulant likely to be chosen because its closed cell structure gives it low density-to-strength properties and makes it very water resistant. CFCs have been used in the past as blowing agents for XPS foam and other closed cell foams and although most manufacturers have now switched to HCFC blowing agents these still have a small ozone depletion potential.
	Ordinary glazing is a poor insulant when compared with a reasonably well insulated wall or roof although multiple glazing systems can have very good insulation values. It is important therefore to optimise window sizes for good natural lighting and only provide larger glazed areas in the right place, where they can benefit from solar gain without giving rise to excessive heat loss or heat gain. Glazing, window frames and surrounding junctions can be cold bridges and thought needs to be given to ensuring there are thermal breaks at these locations. The use of transparent insulation material (TIM) (a honeycomb structure of transparent polycarbonate) fixed in front of a mass wall is another insulation technique which may be appropriate for long south facing walls. Solar radiation passing through glass and the TIM heats the wall behind it, which then transmits stored heat to the interior. The insulation material reduces heat flow back to the exterior. Indirect radiation from an overcast sky can be sufficient to provide useful heat.
▸ D4.3.3 ▸ D4.7.4	More information is given about the selection of insulating materials in the Detailed Design Stage (see D4.3.3) and on avoidance of CFCs in D4.7.4.

| *Background references:* | • Vale, B., Vale, R., *Towards a Green Architecture*, Six Practical Case Studies, RIBA Publications, 1991.
• Hall, K and Warm, P., *Greener Building Products & Services Directory*, Association for Environment Conscious Building Directory, Second Edition, 1993.
• *CFCs in Buildings*, BRE Digest 358 - October 1992 and addenda February 1993. |

| **Good Practice:** | *When considering the appropriate choice of insulants:*
• location and thickness of insulation materials need to be reviewed during the primary design stage when considering the approach to energy conservation, and this will effect the subsequent design detailing, for example whether to use cavity insulation or internal or external applications;
• the designer should be aware of the characteristic and limitations of the insulation materials as this will affect both primary design and detail design:
• the design of construction elements should permit the use of insulants which are CFC free and preferably HCFC free (see Table 5 of BRE Digest 358 for useful guidance);
• optimise window sizes for good natural lighting and limit larger glazed areas to locations where they can benefit from useful solar gain;
• consider the use of insulating windows and provision of thermal breaks at openings and surrounding junctions;
• the use of transparent insulation may be an appropriate technique for buildings with long south-facing walls;
• consider the influence of your choice on fire precautions and particularly on controlling the spread of flame. |

| *Good practice references and further reading:* | • As background references, plus:
• *Window Design*, CIBSE Application Manual AM2, Chartered Institution of Building Services Engineers, 1987.
• CIBSE Guide, Volume A: *Design Data*, Chartered Institution of Building Services Engineers, in sections, 1982–1991.
• *Flat roofing design and good practice*, Chapter 17, Thermal insulants, CIRIA\British Flat Roofing Council, 1993. |

D3.13 Use of materials

Issue D3.13.1	Waste, salvage, re-use and recycling of materials

Background:	The built environment is associated with waste production in a number of ways. Davis, Langdon and Everest have identified the following recycling opportunities in the engineering and civil building sector: • construction waste; • buildings or structures' refurbishment; • waste from operations; • derelict or contaminated land; • materials or components from demolition. There are already well-developed markets for many recycled materials, including ferrous and non-ferrous metals from construction waste. In addition the use of second-hand bricks, roof tiles and slates is widespread. Old doors, fireplaces and other interior items are also now being salvaged and re-used. Recycling of demolition materials by sorting and use of bulk materials as fill is now possible in civil and building construction. Tests carried out in the UK and elsewhere have demonstrated that recycled road planings can be reused in road construction with no appreciable loss of performance. There are practical schemes in Holland and Germany for full sorting, re-processing and re-use of demolition waste. Certain building materials are manufactured from waste in preference to virgin materials. The use of small wood particles in chipboards, blockboard and fibreboards which can be used as substitutes for timber is one example. The use of pulverised fuel ash for concrete block manufacture is another. Designers can promote the recycling of existing buildings and works for re-use as an alternative to new build. This can be a very cost-effective option. Local authorities have a particularly important role to play here through the planning mechanism. Building on brown-field sites, whilst entailing considerable problems (especially if sites are contaminated), can benefit from available public transport infrastructure, rejuvenate derelict land, and displace the need for construction in rural areas.
▶ D3.3.1	

Background references:	• Vale, B. and Vale, R., *Green Architecture – Design for a sustainable future*, Thames and Hudson, London 1991. • Construction Industry Environmental Forum, *Environmental issues in construction – A review of issues and initiatives relevant to the building, construction and related industries*, CIRIA Special Publications 93 and 94, 1993. – Section 4. Resources, Waste and Recycling. • *Waste Recycling and Environment Directory*, Thomas Telford, London 1993. • Contact the University of Surrey Demolition and Recycling Centre, Tel: 0483 300800.

Good Practice:	*To promote recycling and re-use:* • consider building on brownfield sites and refurbishment of existing buildings; • for new buildings, ensure longevity by designing for quality, flexibility and adaptability, low energy use, simplicity and robustness; • design for demountability of the fabric and structure to facilitate recycling; • consider the use of second-hand facing materials such as bricks and roof tiles/slates; • consider drawing up a specification that actively allows for the use of recycled materials and components; • consider directing contractors to the use of products made from recycled materials; • do not be over-ambitious in the quest for increased recycling and recognise when the use of new materials is essential to the design; • recognise that the inappropriate use of recycled materials and products will not promote the long term use of recycled materials; • on civil engineering schemes investigate recycling of bulk materials for fill and road base construction. Major constraints on greater recycling relate to transport costs compared with the value of the materials, cost barriers, and concerns about the durability and performance of re-used materials. *Continued overleaf:*

Good Practice continued:	It must also be recognised that civil engineering schemes are 'one-off' projects. This reduces the opportunity for waste minimisation through standardisation. On the other hand, the commercial success of such projects is often crucially dependent on efficient materials handling and waste reduction.
Good practice references and further reading:	• As background references, plus: • Construction Industry Environmental Forum, *Recycling on site – the practicalities*, Notes of a meeting held on 22 June 1993, CIRIA, 1993. • Hall, K and Warm, P., *Greener Building Products and Services Directory*, Association for Environment Conscious Building Directory, Second Edition, 1993. • Skoyles, E.R. and Skoyles, J.R., *Waste Prevention on Site*, Mitchell Publishing, 1987.

Issue D3.13.2	Design to minimise use of materials
Background:	Over-consumption of construction resources can arise in a number of ways. • Developers, clients and designers can be cautious in specifying their requirements, which can lead to unnecessary use of materials and, in the case of services installations, over-sized plant. • Designers are often cautious, allowing generous safety margins in their design, again resulting in unnecessary use of materials. However this has to be balanced against the flexibility and adaptability which over-design can provide. • Construction products are often not produced in coordinated sizes. A more consistent range of sizes for many materials and components could help to reduce wastage without compromising design flexibility. One simple example of this is the size of standard joinery components which do not co-ordinate with brick-sized openings in external walls. • Lack of attention to design for buildability can result in unnecessary use of resources.
Background reference:	• Howard, N., and Meikle, J., Resources, waste and recycling, Section 4 in *Environmental issues in construction – A review of issues and initiatives relevant to the building, construction and related industries*, CIRIA Special Publications 93 and 94, 1993.
Good Practice:	*These aspects of over-consumption of resources are particularly difficult to deal with, but the designer should consider the following:* • working closely with clients to define their requirements at each stage in the project process in order to match design to need; • ensuring the design and specification is appropriate for the end use of the construction project, for instance in determining floor loadings or works capacity requirements; • striking a balance between over-design and the requirements of flexibility and adaptability, to avoid premature obsolescence of the works or building; • choosing and designing components which can be coordinated, and paying attention to buildability in the design conception; • selecting lightweight demountable methods of construction and materials where appropriate; • working closely with suppliers of construction products to minimise wastage, for example supply of non-standard panel sizes to avoid cutting standard panels.
Good practice references and further reading:	As background reference.

Issue D3.13.3	Storage of recyclable materials
Background:	Even with the most effective steps being taken to minimise waste, building occupants generate paper, cardboard and packaging materials in day-to-day building use. The operation of civil engineering works, for example water or sewage treatment plants, will also generate similar waste, much of which can be re-used or recycled. Some of the waste, such as packaging of chemical additives to the process, will be unsuitable and must be treated as waste for disposal.
	Many households also currently collect reusable materials. Wastepaper, cardboard, glass and metal cans are the most commonly recycled materials and many local authorities now provide collection centres.
	To encourage users of industrial works, building occupants and householders to make efficient use of recycling facilities, adequate storage space, with appropriate fire protection and with access for collection, should be provided. The provision of separate bins in a screened enclosure would help those generating waste to separate and store waste materials prior to having them collected or taking them to a waste collection centre. Retail centres and superstores provide ideal sites for bottle, can and newspaper banks.
	Provision for composting household organic wastes should also be considered at the design stage of residential developments.
Background references:	• Sections on recycling materials in: Prior, J.J., (Editor), BREEAM 1/93, *An environmental assessment for new offices*, BRE, 1993. Crisp, V.H.C, Doggart, J., and Attenborough, M., BREEAM 2/91, *An environmental assessment for new superstores and supermarkets*, BRE, 1991. Prior, J.J., Raw, G.J. and Charlesworth, J.L., BREEAM 3/91, *An environmental assessment for new homes*, BRE, 1991. Baldwin, R., Bartlett, P., Leach, S.J. and Attenborough, M., BREEAM 4/93, *An environmental assessment for existing office buildings*, BRE, 1993.
Good Practice:	***In designing new buildings, industrial premises or civil engineering works:***
	• try to establish from the client whether facilities should be provided for recycling, sorting and collection at the operational stage (in many process-based schemes, this will be an integral part of the project);
	• provide adequate storage space, for example 2 m^2 of storage space/1000 m^2 of floor area in offices and superstores/supermarkets, and an additional 4 m^2 of storage space for each 1000 m^2 of floor area in excess of 2000 m^2 in food and retail stores;
	• in industrial works, ensure such storage facilities are provided with suitable access for collection by truck;
	• provide space for a multiple recycling facility on sites for retail centres and superstores;
	• provide a set of four containers for household waste in a screened and paved enclosure with suitable access for collection in dwellings, also consider space provision for composting household organic wastes.
Good practice references and further reading:	• As background references. • Skoyles, E.R. & Skoyles, J.R., *Waste Prevention on Site*, Mitchell Publishing, 1987.

Issue D3.13.4	Design for maintenance and cleaning
Background:	There are a number of environmental concerns relating to maintenance of civil engineering works and buildings which need to be considered at the design stage.
	In buildings, inadequate maintenance of ventilation and air-handling equipment leads, amongst other things, to poor indoor air quality. Poorly maintained air handling systems release odours and fouling of cooling coils, moisture eliminators and spray ponds by birds and other pests can be a particular problem. Maintenance operations can produce dust and fumes in many ways, for instance in removing panels or tiles for service access. Buildings should, wherever practicable, be designed with access panels for such services so that emissions of fumes, dust gases and vapours are minimised during maintenance and design for maintainability should be actively covered during the design process. Keeping buildings and their occupants in good health also requires attention to design to protect a building from pests.
	In civil engineering, design for ease of maintenance needs to be applied as a principle to a very wide range of situations, for example: • providing access to electrical and mechanical equipment in, for example, water and sewage treatment works, pumping stations and power stations; • building-in access systems to tall and/or long-span bridges; • providing access and/or mechanical debris-clearing devices for river weirs.
	Implications of poor maintenance may be: • increased energy consumption; • pollution through failure of pollution prevention systems; • premature abandonment and/or replacement; • noise nuisance; • reduced safety.
	In both sectors, the choice of facing materials for durability and low maintenance is important for many clients. The selection of external finishes which require high maintenance can result in neglect, unsightly appearance and premature failure of a component or the fabric of the works or building.
Background references:	• References to Maintenance in *Buildings and Health*, see below. • Health & Safety Commission, *Proposals for Construction (Design and Management) Regulations and Approved Code of Practice*, Health and Safety Executive, 1992.
Good Practice:	*Pay attention to the following aspects of design for maintenance and cleaning:* • specifically consider design for ease of maintenance and cleaning early in the design; • electrical and mechanical equipment, including ventilation systems in buildings, should be designed to be easily cleaned, contain no inaccessible surfaces and have surface areas where chemicals can be trapped or absorbed as limited as possible; • ensure services are routed through spaces which can be easily accessed for maintenance and repair; • ensure that appropriate filters are installed in air-conditioning and mechanical ventilation systems and that they are accessible; • consider design to protect a works or building from pests by siting any refuse compounds well away from the main development, limiting unnecessary voids within a works or building, and sealing entry points through the structure fabric; • the durability and maintenance requirements of the structure fabric and external materials and components, bearing in mind that component life and maintenance are a major consideration for most owners of civil engineering works and buildings; • if high maintenance external finishes or components are selected for appearance, ensure the client is fully aware of the maintenance requirements to prevent early deterioration and premature failure and provide suitable access facilities for all exterior parts of the structure.
Good practice references and further reading:	• Curwell, S.R., March, C.G. and Venables, R.K., (Eds), *Buildings and Health, The Rosehaugh Guide to the Design, Construction, Use and Management of Buildings*, RIBA Publications, 1990. See pages 300–304 for a checklist on protecting from pests. • HAPM Component Life Manual, E & F N Spon, 1992.

Stage D4 Detailed design, working drawings and specifications

Stage D4: Detailed design, working drawings and specifications covers the development of the scheme design to detail design by considering the type of construction, quality of materials and standard of workmanship. It includes cost checks and applications for approvals under building regulations, other regulations and to statutory authorities. It also includes the preparation of detailed drawings, schedules and specification of materials and workmanship.

D4.1   Ensuring the design team knows the project environmental policy

D4.2   Legislation and policy

D4.3   Use of energy

D4.4   Labelling and other environmental information

D4.5   Landscape, ecology and the use of plants

D4.6   Internal environment

D4.7   Materials

D4.8   End-use considerations

D4.9   Handover of project environmental policy to contractor and/or other consultants

D4.1 Ensuring the design team knows the project environmental policy

Issue D4.1	Ensuring the design team knows the project environmental policy
Background:	Any initial commitment to produce a design for a project which responds to environmental concerns at every stage can be lost if there are any misunderstandings or failures in communication which lead a member of the team to produce an inappropriate design. It is therefore vital that all members of the team are aware of the project environmental policy and of revisions to it as the design progresses.
Background references:	• Halliday, S.P., *Environmental Code of Practice for Buildings and Their Services*, BSRIA, 1994. • European Construction Institute, *Total project management of construction safety, health and environment*, Thomas Telford, 1992. • Construction Industry Environmental Forum, *Environmental Management in the Construction Industry*, Notes of meeting held on 22/09/92, CIRIA, 1992.
Good Practice:	One member of the team should be appointed to ensure implementation of, but not sole responsibility for, environmental policy and to co-ordinate the environmental aspects of design. ***Each member of the design team should:*** • have a copy of the project environmental policy and be aware of its implications; • receive full and regular briefings on all environmental aspects of the development; • be encouraged to maintain close liaison with outside consultants and where necessary be present at meetings to explain specific issues; • attend team meetings to agree on key issues by presentation of all relevant information; • endeavour to keep abreast of changes in legislation, new information, environmental trends and to inform their colleagues of relevant matters; • acknowledge and try to take account of the concerns of all parties with an interest in the project; • seek appropriate training where issues are unclear or not readily understood. Agreement on team working, freedom of information and clarity in working methods should be achieved.
Good practice references and further reading:	• See background references, plus: • Miller, S., *Going Green*, JT Design Build, Bristol, undated. • Barwise, J., and Battersby, S., *Environmental Training*, Croner Publications, 1993.

D4.2 Legislation and policy

Issue D4.2.1	Building Regulations (As D3.2.2 but repeated here for ease of use)
Current legal position:	The Building Act 1984 is the statutory framework for building control, although the Building Regulations 1991 contain the detailed requirements and procedures. The current regulations, which became effective on 1 June 1992, impose requirements for carrying out certain building operations including the erection of new buildings or the making of a material change of use. The main requirement is that building work must be carried out in accordance with the technical requirements set out in Schedule 1. Approved Documents give guidance on ways of meeting the requirements. There are four main technical requirements of an environmental nature. • ***Precautions against harm to health caused by substances on or in ground to be covered by the building*** – Approved Document C lists types of sites likely to contain contaminants and possible contaminants and actions.

Current legal position continued:	• ***Precaution against the permeation of toxic fumes from insulating material in to any building occupied by people*** – Approved Document D relates this requirement to risks to health of persons from formaldehyde fumes given off by urea formaldehyde foams. If these foams are to be used there must be a continuous barrier to minimise, as far as practicable, the passage of fumes to occupiable parts of the building. The Approved Document gives technical solutions to the requirement including suitability of walls for foam filling, and installation in accordance with BS5618. • ***Adequate means of storage and access to solid waste*** – Provisions to meet compliance with the requirement, contained in Approved Document H, include the capacity, design and siting issues of solid waste storage. • ***Provision for the conservation of fuel and power in certain buildings*** – Approved Document L includes guidance on the limitation of heat loss through building fabric, controls for space heating and hot water supply systems and insulation of hot water storage vessels, pipes and ducts. Resistance to the passage of sound (Part E), ventilation (Part F) and heat-producing appliances (Part J) could also be considered. In complying with the technical requirements there is a further general requirement which provides that 'proper materials' appropriate for the circumstances are used and they are used in a workmanlike manner. Proper materials include those approved under the EC Construction Products Directive and the UK regulations implementing that Directive or a British Standard or Agrément Certificate. Exemptions from the requirements are set out in Schedule 2 of the regulations which include buildings on a construction site which are intended to be used only during the course of those works. There are additional obligations to notify and supply plans to the Building Control Authority (Local Authority or Approved Inspector) when building work is intended to be carried out, and notification of the commencement of building work. There is also a large body of other legislation which may include building control measures. Therefore developers must check the local legislation in force in the area of the construction site to ensure compliance. The regulations do not at present cover certain important environmental areas such as heating efficiency, artificial lighting, ventilation, insulation foams. Overall the generality of the requirements limits their impact in compelling the use of safer building materials.

References to the current legal position:	**EC**• Directive on Construction Products 89/109/EEC.
	UK• Building Act 1984.
	  • Building (Scotland) Act 1970.
	  • Building Regulations 1991 (SI 1991 No 2768) (as amended by SI 1992 No 1180).
	  • Construction Products Regulations 1991 (SI 1991 No. 1620).

Policy and forthcoming legislation:	Each technical requirement outlined in Schedule 1 is supported by a guidance document produced by the DoE – the 'Approved Documents'. Of particular concern are Approved Documents C, D, H and L and that relating to Regulation 7 which relate to the requirements mentioned above. The DoE have consulted on proposals to amend Part L of the Building Regulations and Approved Document L so as to strengthen the requirements for the conservation of fuel and power. This review is in response to the Government's commitment, made at the 1992 Rio Earth Summit, to reduce carbon dioxide emissions. The proposals include the provision of insulation of both doors and windows, draught stripping of doors and windows, improved controls on domestic heating systems and efficient lighting with adequate controls. The DoE also invited views on where air conditioning and mechanical ventilation systems, which are energy expensive, should be allowed and is proposing a method of minimising the use of air conditioning. A floor area of 500 m² is proposed as the lower limit to which the revised regulations should apply.
Policy references:	• DoE, *Approved Documents to the Building Regulations*, 1985, 1990 and 1992. • British Board of Agrément, *Supplement to the Approved Documents* (updated quarterly). • Building Regulation Consultancy Service, Guidance Document, 1992. • DoE, *Energy Conservation: Proposed Amendments to Part L of the Building Regulations 1991*, 1993.

Issue D4.2.2	Avoidance of hazardous materials
Current legal position: ▸ C2.1.6 ▸ C2.1.7 ▸ D3.2.1	Hazardous materials found in building materials and products used in construction are increasingly the subject of legislative requirements. However, there are few products or substances which are currently prohibited from use by law. The supply and use of materials which are forbidden include the most dangerous forms of asbestos products, PCBs and PCTs. Other substances are prohibited for certain uses e.g. the use of PCP-treated wood except in structural timbers inside buildings and lead carbonate and sulphate in paint except in connection with the restoration of historic buildings. These substances are prohibited under EC Directives on the Marketing and Use of Certain Dangerous Substances and Preparations. It is under these Directives that the various UK 'Injurious Substances' regulations are enacted. Following the Montreal Protocol on Substances that Deplete the Ozone Layer the EC has made a Regulation (594/91 as amended by 3952/92) which restricts the production and use of products containing CFCs. Health and safety regulations, in particular the COSHH Regulations, protect the health of persons at work who may be dealing with hazardous materials and any others who may be affected by work using hazardous substances. The Building Regulations 1991 have a general requirement that any building work shall be carried out with proper materials and in a workmanlike manner. Proper materials include those approved under the EC Directive on Construction Products, British Standards or Agrément Certificates. The regulations also include precautions in connection with the use of urea formaldehyde cavity insulation and the presence of toxic substances and contaminant gases. The Planning (Hazardous Substances) Act 1990 and regulations require that hazardous substances consent is obtained from the local planning authority if one or more of the 79 toxic, explosive or flammable substances are present on site in excess of certain thresholds. Threshold values are generally set high, sometimes in the order of hundreds of tonnes of substance on site. The EC Directive on Construction Products provides that construction products (other than listed products which play a minor part with respect to health and safety) must have such characteristics when in use (when incorporated, assembled, applied or installed in the proper manner) to satisfy specified 'essential requirements'. The construction work must be designed or built so that there is no threat to hygiene or health of both occupants or neighbours from certain hazardous substances in the environment. This includes toxic gas, dangerous particles or gases in the air, radiation, pollution or poisoning of water or soil, dampness or the faulty elimination of waste water, smoke, solid or liquid wastes. The Directive is not yet fully in effect since harmonised standards have not yet been developed. However, the end aim is the formulation of such harmonised standards for construction products and the transposition of these standards into relevant national standards. The Environmental Protection (Controls on Injurious Substances) (No. 2) Regulations 1993 have recently been enacted. They restrict the use of cadmium in pigments used in plastic products, and urea formaldehyde, and the use of cadmium in paints containing zinc.

Issue D4.2.2	Avoidance of hazardous materials

References to the current legal position:	**EC** • Directive on Construction Products (89/106/EEC). • Directive on Marketing and Use of Certain Dangerous Substances 76/769/EEC (as amended). • Regulation on Substances that Deplete the Ozone Layer (594/91, as amended). **UK** • Asbestos (Licensing) Regulations 1983 (SI 1983 No. 1649). • Asbestos (Prohibitions) Regulations 1992 (SI 1992 No. 3067). • Construction Products Regulations 1991 (SI 1991 No. 1620). • Control of Asbestos at Work Regulations 1987 (SI 1987 No. 2115) (as amended by SI 1992 No. 3068). • Control of Lead at Work Regulations 1980 (SI 1980 No. 1248). • Control of Pollution (Supply and Use of Injurious Substances) Regulations 1986 (SI 1986 No. 902). • Control of Substances Hazardous to Health Regulations 1988 (SI 1988 No. 1657) (as amended by SI 1990 No. 2026 and SI 1992 No. 2382). • Environmental Protection (Controls on Injurious Substances) Regulations 1992 (SI 1992 No. 31) and No. 2 Regulations (SI 1992 No. 1583). • Environmental Protection (Controls on Injurious Substances) (No. 2) Regulations 1993 (SI 1993 No. 1643). • Environmental Protection (Control of Injurious Substances) Regulations 1993 (SI 1993 No.1). • Health and Safety at Work etc. Act 1974. • Planning (Hazardous Substances) Act 1990 or Town and Country Planning (Scotland) Act 1972. • Planning (Hazardous Substances) Regulations 1992 (SI 1992 No. 656).
Policy and forthcoming legislation:	The EC parent Directive on Marketing and Use (79/769/EEC) is subject to periodic amendments to the Annexes which operate to ban or restrict the use of additional substances. A current proposal (Com(92)195) would restrict certain kinds of creosote for wood treatment. The EC Regulation on Substances that Deplete the Ozone Layer is due to be amended to include methyl bromide and HCFCs. There are proposed Construction (Design and Management) Regulations and a draft Code of Practice. These regulations, if enacted in their current form, would impose requirements on designers to ensure, so far as reasonably practicable, that the design of a building will not expose persons building, maintaining or repairing the structure to risks to their health and safety.
Policy references:	• Government White Paper, *This Common Inheritance – Britain's Environmental Strategy*, 1990. • Government White Paper, *This Common Inheritance – Britain's Environmental Strategy – Second Year Report*, 1992. • DoE Circular 3/92, *Environmental Protection (Control of Injurious Substances) Regulations*, 1992. • DoE Circular 4/93. *Environmental Protection (Control of Injurious Substances) Regulations* 1993. • DoE Waste Management Paper No 6, *Polychlorinated Biphenyl Waste*, 1976. • DoE, *Environmental Action Guide for Building and Purchasing Managers*, 1992. • HSE Approved Code of Practice, *The Control of Asbestos at Work*, 2nd Edition 1993. • HSC Approved Code of Practice, *Work with Asbestos Insulation, Asbestos Coating*, Asbestos Insulating Board, 2nd Edition 1993. • HSC Approved Code of Practice, *Control of Substances Hazardous to Health and Control of Carcinogenic Substances*, 4th Edition 1993.

Issue D4.2.3	Radon and thoron – legal issues
Current legal position: ▸ D3.2.2	There is a requirement under the Building Regulations 1991 (see D3.2.2) to take precautions to avoid danger to health caused by substances found on or in the ground to be covered by a building. Approved Document C recognises that radon is a problem substance and that measures should be taken in areas susceptible to radon. There are no particular legal requirements in respect of thoron but the same principles apply as for radon. Methane should be treated in the same way as gases arising from landfill. The regulations recommend that contaminant gases are removed although in some cases, passive measures alone would be sufficient. The EC recommends a level of 400 becquerels for existing buildings and 200 becquerels for new buildings. The National Radiological Protection Board (NRPB) advises the Government in relation to protection against radiation hazards. Limitation of radon levels in homes is on a non-statutory basis. An advisory 'action level' of 200 becquerels per cubic metre, has been set for new buildings. There are construction standards for new dwellings which are intended to achieve radon concentrations as low as reasonably practicable. Appropriate action should be taken where radon gas is found to be present. Levels of radon are of most importance in the areas (called Radon Affected Areas) identified as having a high incidence of radon namely Devon and Cornwall and parts of Derbyshire, Northamptonshire, Somerset and Yorkshire. The DoE may extend or add to the boundaries of these Radon Affected Areas from time to time. The only context in which there are mandatory rules for the control of radon is in the workplace. Radon produces natural ionising radiation and the Ionising Radiations Regulations 1985 require employers to ensure that exposure to radon is as low as reasonably practicable and within certain limits.
References to the current legal position:	**EC** • Recommendation on Protection of the Public against Indoor Exposure to Radon (21.02.90). • Directive on Protection of the General Public and Workers against Ionising Radiation 80/836/Euratom (as amended by 84/467/Euratom). **UK** • Health and Safety at Work etc. Act 1974. • Ionising Radiations Regulations 1985 (SI 1985 No. 1333). • Building Act 1984 or Building (Scotland) Act 1970. • Building Regulations 1991 (SI 1991 No. 2768) (as amended by SI 1992 No. 1180).
Policy and forthcoming legislation:	In respect of dwelling houses the BRE have issued guidance on inexpensive protective measures which can be installed at the time of construction, to control exposure to radon. BRE advice includes additional sealing of the floor and service entries; and the installation of a radon sump or ventilated underfloor void – see D4.2.4.
Policy references:	• House of Commons Environment Committee, *Report on Indoor Pollution*, 1991. • DoE, *Approved Document C to the Building Regulations*, 1992. • DoE, *The Householder's Guide to Radon*, 1992. • HSE, *A Framework for the Restriction of Occupational Exposure to Ionising Radiation*, HS(G) 91, 1992. • HSE Approved Code of Practice, *The Protection of Persons Against Ionising Radiation Arising from Work Activity*, Part 3, Exposure to Radon, 1988. • HSE, *Radon: in the Workplace*, 1992. • BRE, *Radon: Guidance on Protective Measures for New Dwellings*, 1991. • BRE, *Construction of new buildings on gas-contaminated land*, BR 212, 1991.

Issue D4.2.4	Radon, thoron and naturally-occurring methane – technical issues

Background: ▸ D2.2 ▸ D2.3.1 ▸ D2.3.6 ▸ D3.3.1	These are three naturally occurring gases which tend to seep from the ground and enter buildings through gaps in the floor structure or through open windows and doors. Radon is a radioactive, colourless and odourless gas and exposure to high levels for long periods increases the risk of developing lung cancer. Some regions of the UK have higher natural levels than elsewhere, in particular most of Cornwall, parts of Devon and parts of Somerset, Northamptonshire, Derbyshire and Yorkshire. Work to define the affected areas more precisely is continuing, sponsored by the DoE. Thoron is an isotope of radon and can be treated as a similar gas, although it is less significant. It does not diffuse far from its point of origin before there is significant decay so the radiation doses that people receive are about 1/10th of those from radon. Methane is an asphyxiant, will burn, and can explode in air. It is often associated with landfill sites, where it is generated by the action of anaerobic micro-organisms on biodegradable material. It can migrate into buildings under pressure through subsoil, cracks and fissures. The Building Regulations 1991 from England and Wales require that 'precautions shall be taken to avoid danger to health caused by substances found on or under the ground to be covered by the building.' (Requirement C2). The Approved Document includes in the substances described 'any substance which is or could become ....radioactive'. In 1988 the DoE issued guidance recognising that radon was such a substance and that precautions should be taken in the areas affected.
Background references:	• See good practice references and D4.2.3. • Curwell, S.R., March, C.G. and Venables, R.K., (Eds), *Buildings and Health, The Rosehaugh Guide* RIBA Publications, 1990, Chapter C4.

Good Practice:
 	
▸ D2.3.5
 	
▸ D3.3.1
	• Review D2.3.5 and D3.3.1 and undertake any necessary further soil investigation to determine what, if any, protective measures are necessary. If any signs of possible contaminants are present, contact the local Environmental Health Officer.
	• For existing buildings, the DoE currently fund a free measurement service in areas affected by radon and owners should apply to NRPB. If the concentration of radon is found to exceed a certain level, the 'Action Level', remedial action should be taken although not mandatory: the householder decides. The DoE issued *The Householder's Guide to Radon* containing descriptions and drawings of various remedial techniques.
	• For new dwellings, the BRE guidance contains guidance, advice and drawings on measures to be taken when constructing a suspended concrete floor, in-situ concrete floor, suspended timber floor and a stepped foundation. Examples given are guidance for the more common building situations. In other circumstances, alternative means of achieving the same result are also acceptable. The objective is to provide an airtight barrier across the floor area of the building, with provision for natural ventilation under a suspended floor, or means of extracting soil gas from under a solid floor.
	• Although the guidance is given for dwellings, the following principles, and in many cases the detailed designs, can be applied to other types of building.
	  – Airtight membranes should be carefully detailed with properly executed lapping.
	  – Service entries should avoid penetrating the membrane but, if that is unavoidable, any service entry must have an airtight seal.
	  – Soil gas may enter a building inside a service pipe (duct), so seal the gap between the service cable and the pipe duct with a long-life mastic or flexible sealant.
	  – Consult the BRE report on protective measures for new dwellings.
	  – Consult *Radon Guidance Notes* (NHBC, 1993) which is based on the BRE report.
	  – Consult *The Householder's Guide to Radon* (DoE, 1992) for various remedial techniques applicable to existing buildings. Long-term measurements of radon should be carried out to determine whether there is a significant problem.

Good practice references and further reading:	• *Radon: Guidance on protective measures for new dwellings.* BRE Report 211, 1991. • *The EC Scheme and how it will work*, Factsheet No.1, UK Ecolabelling Board, 1992. • NRPB, *Exposure to Radon Dangers in Dwellings*, HMSO, London, 1987. • NRPB Statement, *Limitation of Exposure to Radon in the Home*, HMSO, 1990. • *The Householders Guide to Radon*, Department of the Environment, 1992. • *Radon Guidance Notes*, NHBC, 1993. • Leach, B.A., and Goodger, H.K., *Building on derelict land*, CIRIA SP78, 1991.

Issue D4.2.5	Contaminated land and toxic substances

Current legal position:	The issue of contaminated land is important because of the risks to human health and to building integrity. Legislation in respect of contaminated land is derived from several statutes. Most of that legislation addresses, either directly or indirectly, legal obligations and liabilities arising in respect of contaminated land e.g. under the Environmental Protection Act 1990 (EPA). However, there are no legislative standards set for clean up of contaminated land *per se* although there is government guidance. The following specific liabilities should be considered by a purchaser or developer of land.
	• **Sections 79–82 EPA statutory nuisance** – liability for any accumulation or deposit which is prejudicial to health or a nuisance.
	• **Section 33 EPA waste management licensing provisions** – the treatment of contaminated soil on site, unless it is excepted by regulations, will require a waste management licence.
	• **Section 34 EPA duty of care regarding waste** – there is a duty on all persons who import, produce, carry, keep, treat or dispose of waste to take all the steps which are reasonable in the circumstances to ensure that waste is disposed of safely.
	• **Water Resources Act 1991** – liability may arise under Section 161 for reimbursement of costs incurred by the regulatory authority in respect of works they considered necessary to carry out in order to prevent poisonous, noxious or polluting matters from entering controlled waters.
	• **Occupiers' Liability Acts 1957 and 1984** – there is a duty to take such care as is reasonable in the circumstances to ensure the safety of persons (visitors and trespassers) entering a site. Liability will ensue for the owner or occupier if that duty is broken.
▸ D3.2.1 ▸ D3.2.2	• **Town and country planning legislation** including the Assessment of Environmental Effects Regulations 1988 (SI 1988 No. 1199 as amended) – contaminated soil may be a material consideration for the purpose of granting planning permission. If it is known about at the time of a grant of planning permission it is likely to be the subject of a planning condition or a planning agreement.
	• **Health and safety legislation** – developers will have obligation to workers on site under the Health and Safety at Work etc Act 1974 and the COSHH Regulations.
	• **Building Regulations 1991** – regarding precaution to be taken in order to avoid danger to health caused by substances on or in the ground which is to be covered by a building.
▸ D2.1	Liability for contaminated land can also accrue under common law. Anyone who suffers damage as a result of the presence or the escape of substances from contaminated land may commence a civil action under the common law rules of nuisance, trespass, negligence or the rule in Rylands –v– Fletcher.
References to the current legal position:	• Environmental Protection Act 1990 and related regulations. • Water Resources Act 1991. • Occupiers' Liability Acts 1957 and 1984. • Town and Country Planning Acts and related regulations. • Health and Safety at Work etc Act 1974 and related regulations. • Building Act 1984, Building (Scotland) Act 1970 and the Building Regulations 1991.

Issue D4.2.5	Contaminated land and toxic substances

Policy and forthcoming legislation:	The only Government guidelines are from the Interdepartmental Committee on Redevelopment of Contaminated Land (ICRCL) which acts as an advisory body to the DoE and those wishing to redevelop contaminated land. The ICRCL issues guidance notes on aspects of contaminated land and their remediation. The ICRCL uses the concept of 'trigger concentrations', depending on the intended use of the site (end use) to assist in determining the significance of contamination and to set acceptable concentrations of contaminants for the different end uses. Remedial methods are also outlined. The guidance is at both a general level and more specific levels, for example in relation to development of gasworks, scrap-yards. The recommendations of the ICRCL may assume the status of a legal requirement if incorporated into a planning agreement. Dutch and American soil clean-up standards are sometimes referred to and may be useful as a comparison to the ICRCL guidance notes.
▸ D3.2.2 ▸ D3.2.1	Approved Documents – the guidance issued by the DoE in respect of the Building Regulations 1991 lists types of sites likely to contain contaminants. They also refer to lists of potential contaminants and state the recommended action to be taken in respect of them.
	Under the planning system, development plans may set out policies for the reclamation and use of contaminated land.
	There are several areas of forthcoming legislation of relevance to contaminated land issues:
	• Section 61 EPA – A date has not yet been set for the implementation of this section. When it comes into force the section will allow a waste regulation authority to recoup any clean up costs incurred by it where it has undertaken works necessary to avoid pollution occurring as a closed landfill site. • Although the proposed registers of contaminative uses have been abandoned – the DoE is to conduct a wide ranging interdepartmental review of the problems of land pollution and the use of contaminated land. • Environmental assessment regulations – the classes of development to which the regulations relate may be extended. • Liability for remedying environmental damage – There is a proposal at EC level to impose strict liability for environmental damage howsoever caused. A Green Paper has recently been issued.
Policy references:	• DoE, *Approved Document C* to the 1991 Building Regulations, 1992. • DoE, *Construction of New Buildings on Gas Contaminated Land*, 1991. • DoE Circular 21/87, *Development of Contaminated Land*. • DoE Circular 17/89, *Landfill Sites: Development Control*. • DoE Draft Planning Policy Guidance (PPG 25), *Planning and Pollution Control*, 1992. • ICRCL Guidance Notes 59/83; 17/78; 18/79; 23/79; 42/80; 61/84; 64/85; 70/90 (1970–1990). • HSE, *Protection of Workers and the General Public During the Development of Contaminated Land*. • Institute of Environmental Health Officers, *Development of Contaminated Land – Professional Guidance*, 1989. • BRE, *Concrete in Sulphate-bearing Ground and Groundwaters*, 1981. • BS5930: 1981, *Code of Practice for Site Investigations*, BSI, Milton Keynes, 1981. • BSI, Draft for Development DD175, *Code of Practice for the Identification of Potentially Contaminated Land and its Investigation*, 1988.

Issue D4.2.6	Electromagnetic radiation – legal issues
Current legal position: ▸ D2.3.5	Electromagnetic radiation is not covered by specific environmental legislation. However, the Electromagnetic Compatibility Regulations 1992 are likely to apply to construction plant. These regulations implement the EC Directive on the electromagnetic compatibility of electrical and electronic appliances, and installations containing electrical or electronic components. Such apparatus must not be supplied unless the equipment complies with the requirements of the regulations which includes protection requirements, conformity assessments, declarations of conformity and affixing the CE mark. In addition, there is a general duty that no person shall take into service relevant apparatus unless that apparatus conforms with the protection requirements relating to both immunity from and emissions of electromagnetic disturbance. The protection requirements are that the electromagnetic disturbance generated must not exceed a level allowing other apparatus to operate as intended and the apparatus itself has a level of immunity to electromagnetic disturbance. Whilst the regulations relate largely to manufacture and supply they also impose a 'user requirement'. This means that the construction industry cannot use new apparatus after 31 December 1995 unless that equipment complies with the regulations. The regulations do not apply to: • apparatus taken into service before 28 October 1992 or • apparatus taken into service until 31 December 1995 if it complies with requirements in force at 30 June 1992 (namely the Wireless Telegraphy Acts). To take apparatus into service in contravention of the regulations is a criminal offence.
References to the current legal position:	**EC** • Directive on Electromagnetic Compatibility 89/336/EEC (as amended). **UK** • The Electromagnetic Compatibility Regulations 1992 (SI 1992 No. 2372).
Policy and forthcoming legislation:	At the end of 1992 the European Commission proposed a Directive on the minimum health and safety requirements regarding the exposure of workers to the risks arising from physical agents. The proposed Directive specifically covers electromagnetic fields (but is concerned only with body-warming effects) as one of four physical agents. The proposal contains minimum health and safety requirements and includes provisions for risk assessment and the provision of personal protective equipment. Annexes 3 and 4 deal with 'optical radiation' and 'fields and waves' and set ceilings (the exposure value which gives rise to risks for an unprotected person which must not be exceeded), thresholds (the value towards which implementation of the Directive is geared) and action levels of exposure (the levels above which a specified measure must be taken). The American Conference of Governmental and Industrial Hygienists has developed recommended threshold occupational limit values for occupational exposure to ultraviolet radiation and lasers. These are sometimes referred to in the UK and may offer guidance at a practical level to avoid possible adverse effects. They are also used to set levels in the Directive mentioned below.
Policy references:	**EC** • Draft Directive on Physical Agents Com(92)560 final, 1992. **UK** • NRPB, *Electromagnetic Fields and Risk of Cancer*, Report of an Advisory Group on Non-ionising Radiation, 1992.

Issue D4.2.7	**Electromagnetic radiation – technical issues**
Background:	The National Radiological Protection Board (NRPB) takes the lead in this area in the UK, and an Advisory Group published a Report in March 1992 having reviewed the experimental and epidemiological data related to the question of EMFs and the risk of cancer. The group concluded that the epidemiological findings that had been reviewed provided no firm evidence of a cancer risk, either to children or to adults, from normal levels of power frequency EMFs, radio frequency or microwave radiation. The evidence strongly suggested that these radiations cannot harm genetic material and so cannot induce cancer. The only remote possibility is that they might act as promoters, that is they might increase the growth of potentially malignant cells. However, the evidence for this was considered to be weak and inconclusive, with the least weak evidence relating to brain tumours. In the absence of any unambiguous experimental evidence to suggest that exposure to these EMFs is likely to be carcinogenic, the findings could be regarded only as sufficient to justify formulating a hypothesis for testing by further investigation. The more recent studies issued in Sweden (as well as in Denmark and Norway) have now also been reviewed by the Advisory Group. The conclusion remains that it is impossible to decide at present whether the hazard, if one exists, is due to exposure to EMF or to some chemical associated with the work carried out in industries where exposure to EMFs may have been increased. The group emphasised the need for research into possible occupational hazards to electrical workers based on objective measurement of EMF and other agents with separate examination of the risk of different types of leukaemia. Two residential studies were also reviewed (from Sweden and Denmark). The Group concluded that these studies were well-controlled and substantially better than those that previously reported associations with childhood cancer. However, the new studies report few cases. They do not establish that exposure to EMF is a cause of cancer, although they provide weak evidence to suggest the possibility exists. The risks, if any, however, would be very small.
Background references:	• *Electromagnetic Fields and the Risk of Cancer*, Report of an Advisory Group of Non-ionising Radiation, Documents of the NRPB, Vol.3 No.1, 1992. • *Electromagnetic Fields, A review of the evidence for effects on health*, BRE Report BR206, November 1991. • Swedish studies pinpoint power line cancer link, Article in *New Scientist*, 31 October 1992. • Pearson, D., *The Natural House Book*, Gaia Books, Conran Octopus, London, 1989, Part 2: Elements (p.81) and Part 3: Spaces (p.201).
Good Practice:	*In the house:* • VDUs (and microwave ovens) can be tested for radiation leakage and protective screens for VDUs are of some help and readily available. *In the office:* • The above points are also relevant to VDUs used in offices. • There is evidence that close proximity of high voltage cables (> 1000 volts) can affect sensitive electronic equipment, for example in laboratories. It may be necessary to consider screening such equipment (i.e. with a Faraday cage).
Good practice references and further reading:	• See background references, plus: • Curwell, S.R., March, C.G. and Venables, R.K., (Eds), *Buildings and Health, The Rosehaugh Guide to the Design, Construction, Use and Management of Buildings*, RIBA Publications, 1990, Appendix 3. • Contact NRPB, Chilton, Didcot, Oxon, OX11 0RQ. Tel: 0235 831600. • Contact BRE, Garston, Watford, Hertfordshire, WD2 7JR. Tel: 0923 894040.

Issue D4.2.8	Safety, security and fire

Current legal position:	The issue of safety and security, both in respect of access to the site and of the hazardous and/or polluting substances that are kept there, is an important element in avoiding the risk of environmental liability attaching to a project.
	One of the common causes of environmental detriment from sites is vandalism and theft. Therefore, in order to avoid environmental liability for pollution arising under the Environmental Protection Act 1990 or the Water Resources Act 1991, sites should be adequately protected by secure fences and locked access.
▶ C2.1.1	
▶ C2.1.4	There is a general duty under the Health and Safety at Work etc. Act 1974 that employers ensure, so far as is reasonably practicable, that people not in their employment (i.e. members of the public and children) are not exposed to risks to their health and safety.
	The Occupiers' Liability Acts 1957 and 1984 require such care to be taken as is reasonable in the circumstances for the safety of visitors, and in certain circumstances even trespassers to premises. The owner or occupier may be liable, if that duty is broken and as a result there is damage or injury to persons or goods on the land due to the state of the property or things done or omitted to be done there. This is important during all stages of the construction phase and safety provisions should be planned at the design stage.
	There are also requirements applying to building operations and works of engineering construction under the Construction Regulations 1961–1966. These requirements relate to safety of construction sites as working places and in respect of means of safe access to those sites.
	The COSHH Regulations apply to the control of hazardous substances on a construction site. The requirement is to ensure that the exposure of employees (and anyone else who might be affected) to substances hazardous to health are prevented or, where prevention is not reasonably practicable, are adequately controlled. This extends to the safe and secure storage of hazardous chemicals in order to comply with the regulations.
	The provisions of the Fire Precautions Act 1971, the Fire Safety and Safety of Sports Act 1987 and the Building Regulations 1991 are ostensibly concerned with the safety of life, for example by the provision of the means of an escape from fire and the prevention of the spread of fire.
	However, if dangerous substances are present on site the Dangerous Substances (Notification and Marking of Site) Regulations 1990 may apply. Under these regulations, persons in control of a site where there is more than 25 tonnes of such a substance, must comply with the provisions of the regulation. These provisions include:
	• the notification to the fire authority; • the marking of the site access with safety signs to give adequate warning to firemen, before entering the site in an emergency, that dangerous substances are present; • the provision of safety signs at the on-site locations, where dangerous substances are present.
	The provision of adequate fire-water containment is advisable so that, in the event of fire, strict liability offences, for example under the Water Resources Act 1991 are avoided.
	The draft Fire Precautions (Places of Work) Regulations 1992 promise sweeping changes in UK Fire Regulations but are still under consultation. The proposed Regulations require employers to:
	• identify the hazard and risk; • prepare emergency plans; • prepare means of fire-fighting; • train staff in fire procedures; • keep records of fire-fighting, training and drills.

Issue D4.2.8	Safety, security and fire
References to the current legal position:	• Environmental Protection Act 1990. • Water Resources Act 1991. • Health and Safety at Work etc. Act 1974. • Occupiers' Liability Acts 1957 and 1984. • Construction (Working Places) Regulations 1966 (SI 1966 No. 94) (as amended). • Control of Substances Hazardous to Health Regulations 1988 (SI 1988 No. 1657) (as amended by SI 1990 No. 2026 and SI 1992 No. 2382). • Fire Precautions Act 1971. • Fire Safety and Safety of Places of Sport Act 1987. • Building Regulations 1991. • Dangerous Substances (Notification and Marking of Site) Regulations 1990 (SI 1990 No. 304).
Policy and forthcoming legislation:	The HSE and CONIAC actively encourage architects and other professionals to be more active in health and safety in the construction industry. CIRIA is undertaking a project (RP467) preparing guidance on increasing inherent security of non-residential buildings. There is an ICRCL guidance note on the fire hazards associated with contaminated land including problems for redevelopment and possible remedial measures. There is HSE guidance on assessing fire hazards and how to protect people at risk.
Policy references:	• HSE, *Site Safety and Concrete Construction, A Guide for Small Contractors*, HS(G)46, 1989. • HSE, *Accidents to children on construction sites*, Guidance Note GS7, 1989. • HSE/CONIAC, *The Control of Substances Hazardous to Health in the Construction Industry*, 1990. • ICRCL Guidance Note *Fire Hazards of Contaminated Land*, 61/84, 1986. • HSE, *Assessment of Fire Hazard from Solid Materials and the Precautions Required for their Safe Storage and Use*, 1991.

D4.3   Use of energy

Issue D4.3.1	Heating

Background:	Review D3.3.7 for background concerning passive solar design and D3.7.1, 2 and 5 for energy conservation and efficiency and design for minimal energy use.
▸ D3.3.7	Heating, both in terms of energy use and maintenance, represents the greatest service cost in buildings and can be a significant cost in some civil engineering plants. The design, selection and maintenance of plant and equipment should be made on the basis of life-cycle costs and environmental impact. Space heating constitutes about a half of the total primary energy consumption in buildings.
▸ D3.7.1	
▸ D3.7.2	
▸ D3.7.5	Design to minimise the heating requirement is essential since this will allow reduction in CO_2 emissions which is a direct consequence of fossil fuel combustion. 'There is general agreement among energy analysts and policy makers that energy conservation will be a major part of any global warming control strategy. The main reasons for this are very short pay back times, usually less than 5 years, and low environmental impact. All energy supply options, nuclear power, wind power, tidal power, solar power, biomass, are environmentally contentious. Energy conservation results in reduced environmental impact regardless of the supply strategy that is chosen.' (Curwell et al 1990)

Background references:	• Curwell, S.R., March, C.G. and Venables, R.K., (Eds), *Buildings and Health, The Rosehaugh Guide to the Design, Construction, Use and Management of Buildings*, RIBA Publications, 1990, Chapter D3, p438.

Good Practice:	*In considering the design and selection of heating systems:*
	• select the least polluting source of energy – the most economical fuel currently available is natural gas, which is also the least damaging environmentally (of fossil fuels) followed by oil and solid fuel, whereas electricity is a relatively poor performer because its generation is inefficient (unless from a hydroelectric plant) and it produces a substantial amount of pollution in relation to usable energy;
	• discuss energy use targets and management and maintenance options at the earliest possible stage, and review decisions made;
	• incorporate all measures to reduce energy (heating) requirement, such as passive solar gain, maximum insulation thicknesses, low emissivity double glazing, high thermal mass, ventilation heat recovery and use of daylight control for artificial lighting;
	• use high efficiency plant and equipment which must not be oversized and must be efficient on part and full load;
▸ D4.3.2	• use condensing boilers in housing (approximately 10% more efficient than the best non-condensing boilers with payback times of the order of 5 years in new buildings);
▸ D4.6.4	• design to avoid air conditioning if possible. Refer to D4.3.2 and D4.6.4, D4.6.8 and D4.6.9 for further information on this topic;
▸ D4.6.7	• select the best and most appropriate control system or programmer, to avoid systems running wastefully or unnecessarily since, with the increasing complexity of many buildings, complicated control systems may be required and, if this is the case, good monitoring and management is essential;
▸ D4.6.8	
▸ D4.6.9	• ensure systems are regularly maintained to maximise their efficiency;
	• provide good access to heating plant for maintenance;
	• allow for any mixed mode requirements, but maintain flexibility as far as possible to cater for possible future change of use; (significant energy savings can be made by use of mixed mode designs, by maximising the passive heating and cooling opportunities that are available, so reducing the length of time when direct heating must be applied during the heating seasons;
	• provide well positioned, easy-to-use personal controls alongside automatic controls in preference to over-riding control, aiming for 'manual on with auto off' or standby when possible;
	• size heater emitters (radiators or equivalent) to suit the boiler size. (Generously sized heat emitters with low flow and return temperatures allow condensing boilers to operate in condensing mode for more of the time, with a small increase in annual efficiency. Similar improvements may be achieved by use of an external temperature compensated primary boiler circuit.)

Good Practice continued:	• investigate opportunity to recycle waste heat generated in one part of the building to another part for heating, particularly relevant for certain factory processes where waste heat could be used to heat offices; • investigate combined heat and power (CHP) systems if appropriate, particularly small-scale CHP which may be feasible for individual housing schemes, or blocks of flats. (CO_2 savings approaching 50% over conventional gas central heating are possible).
Good practice references and further reading:	• See background reference, plus: • CIBSE Guide, Volume A: *Design Data*, Chartered Institution of Building Services Engineers, in sections, 1982–1991. • CIBSE Guide, Volume B,: *Installation and Equipment Data*, Chartered Institution of Building Services Engineers, 1986. • CIBSE Guide, Volume C: *Reference Data*, Chartered Institution of Building Services Engineers, in sections, 1975–1986.
Legal reference:	• Building Regulations 1991: Approved Document L: *Conservation of Fuel and Power*.

Issue D4.3.2	Cooling
Background:	Review D3.9.5 on air conditioning/mechanical cooling which should be avoided due to: • the use of harmful refrigerants such as CFCs and HCFCs; • increased energy consumption (which can be two to three times that of a non-air-conditioned building); • reduced opportunities for individual control over internal temperature and humidity; • increased maintenance requirements and costs. There are cases where air conditioning is unavoidable. The following good practice section offers some detailed measures that can be considered to maximise the efficiency of the plant, thereby minimising the fuel (energy) consumption and cost. In most cases, the very lowest energy consumers are naturally ventilated or only partially mechanically ventilated offices. But air conditioned buildings need not necessarily consume a lot more energy – some low energy cost buildings are air conditioned. Refer to the data given in the DoE, Energy Efficiency Office, Best Practice Programme, Energy Consumption Guide Nos.10 and 19. These indicate that there are many options available to the designer such as partial mechanical ventilation or partial air conditioning. The choice is not necessarily one or the other, but may be a combination of measures to suit the design and client requirements.
Background references:	• Curwell, S.R., March, C.G. and Venables, R.K., (Eds), *Buildings and Health, The Rosehaugh Guide to the Design, Construction, Use and Management of Buildings*, RIBA Publications, 1990: Indexed references to air conditioning. • Building Research Establishment, *CFCs in Buildings*, BRE Digest 358, October 1992 and Addenda February 1993. • Energy Efficiency Office, Energy Consumption Guides 10 and 19.
Good Practice: ▸ D3.9.5	• Carry out an environmental appraisal of the scheme design (incorporating any detail design issues) to predict the likely cooling requirement. Match performance standards and operating hours to user needs. At times of low occupancy, mechanical ventilation systems can waste heat by bringing in too much outside air. Varying fresh air needs can be accommodated using air-quality sensors in the exhaust air ducts to alter the minimum proportion of outside air admitted. • Maximise passive cooling through natural ventilation and minimise summer temperature without mechanical cooling. Review Good Practice D3.9.5 for design factors to consider for reducing dependence on air conditioning. • Install solar control blinds to help control overheating of the internal air and reduce the need for refrigeration plant to operate. *Continued overleaf:*

Good Practice continued: ▸ D3.7.2 ▸ D4.6.9 ▸ D4.6.8 ▸ D4.7.4	• Specify good controls to allow the most efficient and economical running of a building. In buildings with internal rooms, such as conference or lecture rooms and in some deep plan offices adjacent to atria for example, independent operating controls should be considered with simple local timed over-rides. •Use equipment that is efficient in heat recovery and designed for low losses. This is particularly important for lighting which can be linked by automatic controls such that it only comes on when external daylight levels drop or is manually put to 'on' but automatically switches off when activity ceases or external light increases. Provide good task lighting at workplaces to reduce the need for high overall light levels and to allow occupants some control over their environment. This saves energy, reduces heat build-up and gives visual contrast to the spaces. • Concentrate heat generating equipment in small areas, thereby (at least) lowering overall cooling load to localised areas. Mainframe computer suites can sometimes account for more than half the entire energy bill. Their air conditioning often runs inefficiently and offers scope for substantial savings through improved control and management. If possible, power supplies to the computer and its air conditioning should be separately metered and regularly monitored. If the ratio of air conditioning to computer consumption is more than 0.6, there may well be scope for improvement. Unified control should be considered for installations made up from independent packaged units. • Heating and ventilation systems should be controlled in zones to maximise flexibility and control, particularly if certain parts of the building are regularly used outside normal hours. • Use maximum insulation ('super-insulation') of the fabric: double or triple glazing and airtight construction to minimise energy use. (Review D3.7.2). • Advise on the need for staff training (if appropriate) and provision of written instructions to ensure the building is properly managed. In some cases a building management system may be worthwhile. In all cases, good maintenance is essential in minimising energy consumption and the client must be made fully aware of this. • If air conditioning is unavoidable, refer to D4.6.9 and D4.6.8 for good practice on system design, installation and maintenance. • Use refrigerants with low or no ozone depletion or global warming potential (see D4.7.4). • Consider carefully the fan power and controls in air conditioning because excess running hours can lead to unseen and undetected waste – an 'hours-run' meter on important items of plant will facilitate monitoring. • Consider provision for ice storage as a means of reducing the chiller size and reducing electrical costs whilst recognising its higher energy usage. • Ensure that all building services equipment and controls are fully commissioned before handover to the client, owner and/or occupier.
Good practice references and further reading:	• Curwell, S.R., March, C.G. and Venables, R.K., (Eds), *Buildings and Health, The Rosehaugh Guide to the Design, Construction, Use and Management of Buildings*, RIBA Publications, 1990 – Pages 443 and 444, and Chapters C3/C5, 1990. • Building Research Establishment, *CFCs in Buildings*, BRE Digest 358, 1992. • Halliday, S.P., *Environmental Code of Practice for Buildings and Their Services*, BSRIA, 1994. • Chartered Institute of Building Services Engineers' Technical Memorandum 13 – *Minimising Risk of Legionnaire's Disease*, London, CIBSE, 1987. • BS6700: 1987 *Specification for Design, Installation, Testing and Maintenance of Services Supplying Water for Domestic Use Within Buildings and their Curtilages*, BSI, Milton Keynes. • *Water Fittings and Materials Directory* 1989, Water Byelaws Advisory Service. • HSE Contract Research Report 42/49. • CIBSE (Chartered Institute of Building Services Engineers): (a) *Environmental Criteria for Design*, Guide Section A1. (b) *Ventilation and Air Conditioning Requirements*, Guide Section B2. • Finnegan, M.J, Pickering, C.A.C., and Burge, P.S., The Sick Building Prevalence Studies, in *British Medical Journal*, 289, 1573–5, 1984. • *Renewable Energy in the UK: the way forward*, Energy Paper No.55, HMSO, London, 1988. • Good Practice Case Studies published by the Energy Efficiency Office (Numbers 13, 14, 15, 17, 18, 19, 20 and 21).

Issue D4.3.3	Insulation

Background:	Insulation is fundamental to any environmental policy for a building both in conserving energy and reducing the amount of CO_2 released into the atmosphere. The type of insulation and its thickness need to be considered. In addition to thermal insulation the choice of insulation material needs to be considered against both fire and sound insulation criteria. The type of glazing can help enormously in improving insulation. In a typical single glazed house, as much as 25% of heat wasted is lost through the glazing. Double glazing can reduce this heat loss by half and double glazing incorporating low emissivity glass will halve the loss again (DoE Calculations). Proposed revisions to the 1991 Building Regulations are likely to encourage the use of double-glazing for all new buildings. For insulation, avoid products that are manufactured using ozone-depleting CFCs as the blowing agents, such as certain types of rigid polyurethane foams, extruded and expanded polystyrene foams and phenolic foams. Use recycled paper, mineral fibres, foamed glass or non-CFC blown expanded polystyrene. At present, recycled paper insulation (arguably the best option) has a British Board of Agrément certificate for use only in roof spaces. The manufacturer claims recycled paper is suitable for insulating walls in timber-framed houses, but it cannot be used to fill cavity walls. It has a tendency to compress when it absorbs water which reduces its thermal resistance, and has to be 'topped up' to maintain the required value. Cellular glass insulation is inert, has minimal effect on the environment and is extremely durable. It maintains its thermal resistance because it does not absorb any water and is easy to dispose of when the building is demolished. It is thought by some environmentalists that mineral fibres (glass and rock fibres) may pose a health risk when the building is demolished, although this is strongly denied by the manufacturers and is not proven. On the plus side, the raw materials used to make mineral fibres come from a plentiful (almost inexhaustible) source and the fibres can be recycled. Assessing the embodied energy of an insulant is often difficult and many manufacturers resist giving data. But by far the best is recycled paper insulation, estimated to have about 1/10th the embodied energy of mineral fibre-based insulation. Mineral fibre insulation is, in turn, generally lower than polymeric insulants such as expanded polystyrene.
Background references:	• Hall, K and Warm, P., *Greener Building Products and Services Directory*, Association for Environment Conscious Building Directory, Second Edition, 1993. • Building Regulations, 1992: Approved Document L. *Conservation of Fuel and Power*.
Good Practice: ▸ D3.8.1 ▸ D3.12.7	The type of insulation selected must be appropriate for its application, and in particular, be durable to suit the life span of the building. Options range between: **1. Superinsulation:** • low fabric U-values in walls, roof and floor • triple glazed windows • airtight construction • controlled ventilation system **2. Standard insulation:** • Building Regulations minimum • passive solar design (large south facing glazing and minimal north facing glazing) • heavy construction (thermal mass). Good insulation standards for walls, floors and roofs are set out in D3.7.2. In addition, use of low emissivity double glazing has a U-value of 1.9W/m°C (against 2.8 for standard double glazing). Choice of insulation must be considered against the following desirable criteria: • to not contain CFC gases, or give off unacceptable pollutants when manufactured; • to have low embodied energy; • should be made from recycled, renewable or readily available raw materials; • to not harm building workers or occupants; • to be durable, but not pose problems from disposal after use.
Good practice references and further reading:	• Vale, B. and Vale, R., *Green Architecture – Design for a sustainable future*, Thames and Hudson, London 1991. • Shades of Green, article in *Building*, 22 January 1993. • BRE, *Thermal Insulation; avoiding risks*, BRE, 1989. • NHBC, *Thermal Insulation and ventilation*. Good Practice Guide, November 1991.

D4.4 Labelling and other environmental information

Issue D4.4.1	Energy labelling schemes
Background: ▸ D3.8.1 ▸ D3.8.3	Refer to D3.8.1 for information and background on the two audit schemes available for residential designers. These are the National Energy Foundation's 'National Home Energy Rating' scheme (NHER) and MVM Starpoint. The Government, in recognising these two assessment methods, has introduced a standard assessment procedure (SAP) which rates homes on an energy efficiency scale of 1 to 100. (This relates approximately to NHER values of 1 to 10). It is likely that the next revision to the Building Regulations Approved Document L will require a SAP rating to demonstrate compliance with the regulation. Similar energy labelling schemes are not currently available for other building types or other types of construction project. Reference should also be made to D3.8.3 for details of the BREEAM scheme (Building Research Establishment Environmental Assessment Method). Finally, a useful source of reference is BRECSU (Building Research Energy Conservation Support Unit) which manages and promotes the Best Practice Programme for all buildings sectors on behalf of the DoE's Energy Efficiency Office (EEO). Whilst not operating a labelling scheme as such, its programme aims to advance energy efficiency in the UK through collaboration with energy consumers and those offering advice, services and techniques.
Background references:	See good practice references.
Good Practice: ▸ D3.8.1	• For residential developments, review D3.8.1 and carry out an energy audit based on the NHER or MVM Starpoint schemes, if not already done. • Aim for an NHER rating of at least 8 or better, bearing in mind that a value of 10 is relatively easy to achieve. This is equivalent to an SAP of 80 or better. The NHER programme allows for substitution of different elements (e.g. thicker insulation, more efficient appliances etc.) allowing quick comparisons to be made, including 'pay-back' periods. • For other building types, at present, there is no equivalent scheme although Part 2 of the CIBSE Building Energy Code can be used to compare options. Develop the detail design having regard to all the principles set out in this and previous sections. • In addition, refer to examples of energy efficient designs for the particular type, for example: *Offices:* – NMB Bank, Amsterdam; – Spectrum 7 Office, Industrial Building, Milton Keynes; – Lansdown Passive Solar Office, Milton Park (not built). – National Farmers Union Mutual and Avon Insurance Group HQ, Stratford on Avon; – Policy Studies Institute, London NW1 (refurbishment); *Health care:* – Woodhouse Medical Centre, Sheffield; – St.Mary's Hospital, Isle of Wight; *Student housing:* – Halls of Residence, Strathclyde University, Glasgow; *Schools:* – Grove Road Primary School, London Borough of Hounslow.

Good Practice continued:	• Carry out a BREEAM Assessment (schemes available for homes, new offices, existing offices and new superstores and supermarkets and one for industrial buildings is due soon).
	• Refer to BRECSU (Building Research Energy Conservation Support Unit) which works within the DoE Energy Efficiency Office framework. This publishes Good Practice Guides and Good Practice Case Studies on various building types, as well as Energy Consumption Guides. Incorporate measures as necessary within the cost allowance.
Good practice references and further reading:	• Atkinson, C.J., and Butlin, R.N, *Ecolabelling of building materials and building products*, BRE Information Paper 11/93, BRE, 1993.
	• Prior, J.J., (Editor), BREEAM 1/93, *An environmental assessment for new offices*, BRE, 1993.
	• Crisp, V.H.C, Doggart, J., and Attenborough, M., BREEAM 2/91, *An environmental assessment for new superstores and supermarkets*, BRE, 1991.
	• Prior, J.J., Raw, G.J. and Charlesworth, J.L., BREEAM 3/91, *An environmental assessment for new homes*, BRE, 1991.
	• Baldwin, R., Bartlett, P., Leach, S.J. and Attenborough, M., BREEAM 4/93, *An environmental assessment for existing office buildings*, BRE, 1993.
	• NEF (NHER), Milton Keynes: 0908 672787.
	• MVM Starpoint, Bristol: 0272 250948.
	• BRECSU (at BRE): 0923 664258.
	• BREEAM (at BRE): 0923 664462.
	• EEO: 071-276 6200.

Issue D4.4.2	**BREEAM**
Background: ▶ D3.8.3	BREEAM (Building Research Establishment Environmental Assessment Method) is fully covered in D3.8.3, and a review of this at this stage is essential if not already undertaken. The purpose of the scheme is to evaluate the environmental effects of the building at design stage. This is more relevant to scheme design and should, therefore, already have been carried out so that improvements can be investigated and incorporated before the design is fixed. However, there are issues of relevance to detail design, so a review or assessment at this stage should be undertaken.
Background references:	See D3.8.3.
Good Practice: ▶ D3.8.2	• See D3.8.3 and arrange for BRE's approved assessors to carry out an assessment if appropriate. (Allow 2 to 3 weeks). • Reconsider any of the more important primary design decisions if necessary (i.e. if the client requires an improved rating and if time and costs permit).
Good practice references and further reading:	See D3.8.3.

D4.5   Landscape, ecology and the use of plants

Issue D4.5.1	Landscape and ecology
Background: ▸ D2.3.3 ▸ D3.6 ▸ D3.3.2	Detailed design and scheduling should be worked up on the basis of decisions made at preliminary design stage. In respect of wildlife areas it may involve careful restoration of existing habitats or creation of new ones. Choice of species, their size, age and provenance will have to be made. Local varieties and plants sourced from nearby to the development are more likely to retain the important features of the site ecological character than those which are not local.
Background references:	See good practice references.
Good Practice:	*To ensure landscape and ecology are adequately covered in your design:* • employ ecological and landscape design expertise where necessary; • consult interested parties – including the NRA if river ecology is an issue – at every stage to seek input to the design and acceptance of proposals; • reflect the composition of local habitats by careful ecological design and selection of local varieties; • where possible, choose materials from local sources.
Good practice references and further reading:	• BS1192: 1984 Part 4: *Recommendations for landscape drawings*, BSI, Milton Keynes. • BS3936: 1980 Part 1: *Nursery Stock – Specification for trees and shrubs*, BSI, Milton Keynes. • BS3875: 1969 Part 5: *Horticultural, arboricultural and forestry practice*, BSI, Milton Keynes. • BS4428: 1979 *Recommendations for general landscape operation*, BSI, Milton Keynes. • BS4043: 1978 *Recommendations for transplanting semi-mature trees*, BSI, Milton Keynes. • Lovejoy & Partners, *Landscape Handbook*, 3rd Edition, E & F N Spon, July 1986. • Clamp, H., (Editor), *Landscape Contract Manual*, E & F N Spon, 1992. • Lovejoy & Partners, *Landscape and External Works Price Book*, 13th Edition, E & F N Spon, October 1993. • Ministry of Agriculture, Fisheries and Food, *Environmental Procedures for Inland Flood Defence Works – A guide for managers and decision makers in the NRA, Internal Drainage Boards and Local Authorities*, 1992.

Issue D4.5.2	Use of plants to 'green' the built environment
Background:	Increasing interest is being shown in using plants on, and in close proximity to, buildings and civil engineering works, not only to provide more green space, but also to improve the overall environment of the immediate area. Roof gardens help to replace green space lost to a development, provide areas for recreation and public enjoyment and help regulate storm water run-off. Green walls can soften an otherwise hard, built landscape and act as a counterpoint to engineering or architectural detail. Both can increase energy efficiency, may help to protect a structure's fabric and to filter out dust and pollutants, whilst helping to oxygenate the air and ameliorating the microclimate. Planting trees and shrubs in close proximity to buildings and within the building can also have a beneficial effect on people and microclimate. Vegetation can also be used to provide or enhance engineering solutions to some problems or design needs. Used in conjunction with geotextiles, vegetation can help, for example, to avoid or reduce erosion, instability, river bank collapse.
Background references:	See good practice references.
Good Practice:	*In considering the use of plants to 'green' your project:* • take opportunities for greening walls and roofs where appropriate, particularly where there are large expanses of overlooked unattractive walling or flat roof or where green space is otherwise restricted; • consider the additional loading on any roof to which you choose to add a layer of soil and plant material; • seek expert advice on the design of waterproofing and irrigation systems and the final planting design and choice of species; • identify the means by which soil and other necessary materials will be moved onto the roof for ongoing management; • ensure your plans take account of the aspect of the particular wall(s) concerned since it will affect the successful establishment of plant material and also its capabilities to improve the energy efficiency of whatever the wall is a part – for example evergreens on a west-facing wall will protect it from rain and provide a cushion of air to help insulation; • design-in the erection of a framework on to which climbing plants can grow; • if creating green areas within the curtilage of a building, ensure that the shape and size of species, particularly trees, to be introduced are appropriate and that the impact their root systems may have on structures and services in close proximity has been evaluated; • consider the use of indoor plants which are capable of absorbing pollutants from the internal atmosphere.
Good practice references and further reading:	• Johnston, J. and Newton J., *Building Green – A guide to using plants on roofs, walls and pavements*, London Ecology Unit, 1993. • Coppin, N.J. and Richards, I.G., *Use of vegetation in civil engineering*, CIRIA Book 10, 1990. • *Protection of river and canal banks*, Book 9, CIRIA, 1990.

D4.6 Internal environment

Issue D4.6.1	Services
Background:	Designers should aim at as naturally-serviced an environment as possible. The services-related issues in the rest of section D4.6 emphasise this in an attempt to reduce the energy and others requirements of a building or works. A thorough review of the services issues is recommended at this stage to confirm the overall services strategy. This is the stage at which to reconsider or redefine the brief for services and ensure that all members of the design teams and especially the client, are aware of the brief and understand how the building will be used and managed.
Background references:	• Halliday, S.P., *Building Services and Environmental issues – The Background*, BSRIA Interim Report, April 1992.
Good Practice: ▸ D4.6.2 – 9	*When considering the services provision for your project:* • review the other issues in the rest of section D4.6, and ensure that any necessary measures are implemented; • prepare a strategic services policy, encapsulating all the principles agreed above and in line with the provisions of the BSRIA code of practice; • ensure that there is a coordinated environmental strategy for incorporation of services design and construction with the architectural and structural work; • consult all the mains services companies and statutory undertakers and agree mains distribution routes, intake details, etc; • ensure that the essential services loads can be met, but avoid over-capacity; • ensure the performance specification defines the most efficient plant and distribution systems, as well as using the most environmentally friendly products; • ensure that the performance specification defines the most appropriate control systems for different zones, and make the client aware of the operational requirements; • ensure that the client is fully aware of all maintenance requirements, and stress the importance of good, regular maintenance; • ensure there is adequate maintenance access, and make provision for replacement and alteration of plant; • undertake a building environmental assessment.
Good practice references and further reading:	• Halliday, S.P., *Environmental Code of Practice for Buildings and Their Services*, BSRIA, 1994. • *Design Briefing Manual*, AG1/90, BSRIA, 1990. • Hejab, M. and Parsloe, C., *Space and weight allowances for building services plant*, Technical Note 9/92, BSRIA, 1993. • Construction Industry Environmental Forum, *Environmental issues in construction – A review of issues and initiatives relevant to the building, construction and related industries*, CIRIA Special Publications 93 and 94, 1993. • *Operating and maintenance manuals for building services installations*, BSRIA AG1, 1987. • *Maintenance Management for Building Services*, CIBSE, TM 17, 1990. • *Building Services Maintenance*, BSRIA, RG No.1, Vol 1 & 2, 1990. • *Standard Maintenance Specification for Mechanical Services in Buildings*, HVCA. • *Decision in Maintenance*, BSRIA, TN14/92, 1992. • CIBSE Guide, Volume A: *Design Data*, Chartered Institution of Building Services Engineers, in sections, 1982–1991. • CIBSE Guide, Volume B: *Installation and Equipment Data*, Chartered Institution of Building Services Engineers, 1986. • Contacts: BSRIA (Building Services Research and Information Association) 0344 426511. BRECSU (Building Research Energy Conservation Support Unit) 0923 664258. CIBSE (Chartered Institution of Building Services Engineers) 081-675 5211.

Issue D4.6.2	Water use and disposal

| **Background:** | Although water supplied in the UK is essentially a safe commodity, the quantity we consume is as important as the quality of our water. Water metering is gradually being introduced and experience suggests that water consumption drops by 40 per cent once meters are installed. Meters are also early warning devices in case leaks develop, although the major water leaks occur in mains supply pipes.

As only about 5–10% of water is used for drinking and cooking it can be worth considering introducing an integrated water conserving system within a building using special rainwater collection, waste water recycling and landscape techniques, to allow re-use of rainwater and wastewater.

It should also be remembered that water is a useful natural moderator of temperature and humidity within a building, and when introduced whether as cascades or fountains can be an attractive feature to give life to an interior. Such features need to be kept clean and treated to avoid *Legionella*. |
|---|---|
| *Background references:* | • Forster, C., *Water Quality* in *Buildings and Health*, RIBA, London, 1990.
• Hall, K and Warm, P., *Greener Building Products and Services Directory*, Association for Environment Conscious Building Directory, Second Edition, 1993.
• Pearson, D., *The Natural House Book*, Conran Octopus, London, 1989. |
| **Good Practice:** | *In designing for water use within a building:*

• agree the strategy for water use storage and disposal, and minimise mechanical services pipework in relation to system efficiency;
• consider installation of point-of-use domestic and commercial water filters at sinks and drinking water outlets, usually of the activated carbon type;
• consider installation of a separate tap for drinking water;
• ensure hot water supply to drinks vending machines dispenses the drinks at a temperature of about 70°C, by ensuring there is no short circuiting of the header tank, the hot water supply is compatible with the demand and there is a safety cut-out to prevent drinks being taken at too low a temperature;
• when refurbishing buildings, strip out all lead pipework for water supplies and specify copper or medium density polyethylene (MDPE) plastic pipes;
• in leisure buildings and public or commercial sector buildings consider using an attractive water feature, possibly in conjunction with an atrium or winter garden, whilst ensuring it is designed to permit regular cleaning and treatment.

To reduce water use:

• specify low water flush toilet and cisterns (6 litres or less), low flush urinals, and water-efficient showers and taps;
• allow for the collection of rainwater from building roofs and surrounding pavings either into a water butt or an underground rainwater cistern; for re-use in gardens, amenity spaces and greenhouses, but note that water from roofs clad in lead or with lead valleys is not suitable;
• consider waste water recycling, (by collecting wastewater from baths, showers and basins (also known as greywater) and filtering it through sand, gravel or biological filters before collecting into an underground cistern, but note this system must be properly maintained if it is not to pose a health risk;
• as a last resort in areas of constant water shortage or erratic water supplies consider the use of the waterless compost toilet in dwellings. |
| *Good practice references and further reading:* | • Hall, K and Warm, P., *Greener Building Products and Services Directory*, AECB, 1993 contains useful information on automatic urinal controls, dry composting toilets, water storage cisterns and domestic and commercial water filters.
• Halliday, S.P., *Environmental Code of Practice for Buildings and Their Services*, BSRIA, 1994.
• *Plumbing Engineering Services Design Guide*, Institute of Plumbing. |

Issue D4.6.3	Noise and acoustics

Background:	Noise (defined as sound which is undesired by the recipient) whether from external or internal sources is of fundamental importance to the building designer and services engineer.
	The Building Regulations 1991 (as amended) include in Part E the requirement for dwellings, that a wall, floor or stair 'shall resist the transmission of airborne (and, in the case of floors, impact) sound'.
	Noise, loud enough (about 150dB) to cause immediate damage to hearing will not normally occur in buildings. The difficulty with noise control is to establish a 'reasonable' level that provides a comfortable environment for building occupants. It is this that the Building Regulations aim to achieve, although to some occupants the standard may not seem to be adequate because individuals vary in their reactions to noise. When it is a question of damage to hearing, or of interference with speech, it is possible to specify within reasonable limits what the effects of noise will be (although there may be some variation between individuals). When it is a question of annoyance then there is a wide range of responses. It is important to note that Part E of the Building Regulations contains sound insulation requirements for separating walls and floors in new-build dwellings and flat conversions. The requirements can be met by adopting one of the forms of construction given in the Approved Document or by demonstrating that a given numerical performance standard has been achieved elsewhere.
▸ D2.3.7	
▸ D3.4.3	The scheme design stage has dealt with strategic issues resulting from external noise sources such as airports, roads, railways or factories. It is at this stage, however, that noise has to be considered in detail, and decisions taken to ensure a satisfactory environment for the buildings occupants.
	It is probable that personal and regulatory standards for noise within buildings will become more stringent, although at present the emphasis is more aimed towards standardising the existing standards and ensuring compliance with them. We have already seen changes to Approved Document E to the Building Regulations which came into force in June 1992, dealing largely with noise between adjoining dwellings, but standards in the UK still fall below those in Germany and France. For general background to UK sound control for homes as well as a detailed review of the current Building Regulations, together with worked examples, refer to *Sound Control for Homes* (see below).
	External noise sources can be one of the major factors that determine the need for air conditioning. Buildings in city centres, or close to airports, motorways or busy traffic routes are clearly examples where natural ventilation by occupants opening windows is simply not acceptable.
▸ D4.3.2 ▸ D4.3.3	Acoustics experts recognise that there are two issues that determine sound control within buildings:
	• setting appropriate standards (i.e. the Building Regulations); • achieving compliance with these standards.
	The attention to detail is crucial to the latter point. The physical characteristics that contribute to sound insulation are mass and isolation, which operate in three main situations (in each, the observer is within the building:
	• room-to-room airborne sound insulation, i.e. the source is within the building; • impact sound insulation; • outside-to-inside airborne sound insulation i.e. the source is outside the building;
	The following good practice section offers a number of practical suggestions for the designer to consider. In specialised building types, such as theatres and concert halls, etc, an acoustics consultant should be appointed to offer expert advice.
Background references:	• Refer to D3.4.3. • *Sound Control for Homes*, Joint Publication by BRE and CIRIA, 1993.

Issue D4.6.3	Noise and acoustics
Good Practice: ▸ D3.4.3	Review D3.4.3 and implement any relevant measures. In particular, review internal organisation of the building or facility as follows. • In residential planning, non-habitable rooms (kitchens, bathrooms, stores and halls) should face any external noise source to act as a barrier. • Try, in non-residential buildings, to place store rooms, service areas and circulation routes around noise sources, such as lifts or computer rooms. • Residential planning must ensure similar spaces adjoin (i.e. kitchens next to kitchens, etc.) in terraced houses and, in blocks of flats, similar spaces must 'stack' vertically to avoid noise nuisance, a feature that is especially important with bedrooms which must adjoin other bedrooms, not kitchens or living rooms of adjoining dwellings. • In high rise buildings, less sensitive rooms like store rooms or kitchens should be located next to sources of noise such as lifts, refuse chutes, common staircases, motor rooms etc.. • Note that trees and hedges are not very effective noise barriers although specially constructed willow walls can be very effective. A tree belt 100 m deep may reduce road traffic noise levels received on the other side by only about 3dB more than the attenuation normally achieved with soft ground. • Select service plant carefully to suit its location. If air conditioning is required, noise levels and transmission through the structure must be controlled, taking into account the activities that will take place in the rooms, with no high pitched or intermittent noise in general working areas. The designer must ensure that an acceptable steady background noise level is specified. • Try to isolate office machinery from activities needing quiet, where possible, especially photocopiers, plotters, laser printers, etc, when they are used in clusters or continuously during the day. Ensure that non-isolated items do not create aggravating sounds. • In detailing the building, refer to Part C of *Sound Control for Homes*. This gives valuable guidance on the selection of appropriate forms of construction to control external and internal noise, as well as for detailing for noise control. Reference should also be made to the worked examples in Part D. This covers the more frequently encountered acoustic problems, and deals with most practical situations, including the design and construction of new dwellings, sound insulation in conversions and noise problems in existing dwellings. • Refer to Guides A1 and B12 published by the Chartered Institution of Building Services Engineers (CIBSE) for detailed advice and guidance for the control of noise from ducted air ventilation systems. • Select plant (particularly fans) that will operate with low noise. • Mechanically isolate fans and other dynamic plant items, from the building structure and use flexible connectors between fans and ductwork, and resilient mountings where appropriate. • Avoid routing ducts through sensitive rooms. • Acoustically seal around ducts, pipes, conduits and trunking where they pass through partitions, separating walls and floors. • Minimise airflow-generated noise by designing simple layouts with radius bends, flared sections and smooth surfaces. • Install in-duct silencers to reduce fan noise, which should be located close to the fan. • Lag ducts to reduce noise transmission through the duct walls. • Refer to CIBSE Guide B12 for noise control mechanisms and criteria for water systems. • Careful location of washing machines in flats is important. The best location is on a solid ground floor. The most troublesome location is on a lightweight upper floor where the resonant frequency of the joists may be near the machine spin frequency, resulting in vibration amplification. If unavoidable, to minimise the effects in practice, locate the washing machine near the end of joists where the deflection is least.
Good practice references and further reading:	• See background references, plus: • CIBSE Guide, Volume A: *Design Data*, Chartered Institution of Building Services Engineers, in sections, 1982–1991. • CIBSE Guide, Volume B: *Installation and Equipment Data*, Chartered Institution of Building Services Engineers, 1986, Section B12, Sound Control.
Legal references:	• Building Regulations 1991: Approved Document E, *Resistance to the Passage of Sound.* • Health and Safety at Work etc Act 1974.

Issue D4.6.4	Ventilation and air quality

Background: ▸ D4.3.2 ▸ D3.9.1 ▸ D3.9.5 ▸ D4.6.8	Issues which adversely affect air quality and guidance on the means of dealing with them are dealt with elsewhere. The emphasis on avoiding air conditioning (see D4.3.2) increasing the 'air-tightness' of the building envelope (see D3.9.1 and D3.9.5) are just two examples of measures taken to reduce energy consumption. None of these measures can be at the expense of indoor air quality which is essential for the health, safety and comfort of building occupants. 'Investigations have found that naturally ventilated buildings in general have worse indoor air quality than air conditioned ones, but the incidence of sick building syndrome is lower' (Paul Appleby et al p.186, in *Buildings and Health*). The most successful buildings are those that achieve energy efficiency in terms of heating and cooling, whilst maintaining a comfortable, healthy environment. In housing, this is easier to achieve by less sophisticated means. In commercial or hospital buildings, more care needs to be taken to achieve good indoor air quality. Two good examples are the NMB Bank, Amsterdam and St.Mary's Hospital, Isle of Wight. In non-residential buildings, there is a need to distinguish between comfort and health in the context of indoor air quality and ventilation requirements. Standards are either comfort-based (e.g. the provision of adequate outdoor air, reducing concentrations of specific contaminants below their detection threshold or reducing total odour loads) or health-based (e.g. reducing concentrations below specific occupational exposure limits). It is essential for the designer to develop a strategy to eliminate or reduce occupational exposure to contaminants. *Buildings and Health*, Chapter C1 sums this up as: • elimination of contaminant; • substitution with less malodorous contaminant as appropriate; • reduction of contaminant emission rate; • segregation of occupants from potential sources of toxic or malodorous contaminants; • provide ventilation at source or dilute to acceptable concentration; • personal protection, e.g. workers wear air-fed respirators. The difficulty for designers is to define what is 'acceptable' in terms of indoor air quality, which therefore means that for a certain proportion of occupants, this standard may not be reached. The widely acknowledged standard i.e. (that applied in the UK and USA) is based on the concept of an acceptable level of dissatisfaction, normally taken as 20%. A comfortable indoor air quality is therefore taken primarily to be one which is not unacceptably malodorous. Some contaminants cannot be detected by normal human senses, but may still cause short or long term health effects. To cater for this, Occupational Exposure Limits provide limits on occupational exposure to airborne substances hazardous to health, and are primarily used for assessing compliance with the UK Health and Safety Legislation. These guidelines are useful in ensuring the correct balance between energy conservation and satisfactory air change rates. However, these only cover a small proportion of the pollutants present in modern buildings. Reference material on indoor contaminants and some common organic compounds found in office atmospheres can be found in *Buildings and Health*, Chapter C1, pp.170–171.
Background references:	• Curwell, S.R., March, C.G. and Venables, R.K., (Eds), *Buildings and Health, The Rosehaugh Guide to the Design, Construction, Use and Management of Buildings*, RIBA Publications, 1990. • *Air Quality Guidelines for Europe*, World Health Organisation, Geneva, 1987. • Control of Substances Hazardous to Health Regulations, 1988. • Building Regulations 1991, Part F. • Workplace (Health, Safety and Welfare) Regulations 1992.

Issue D4.6.4	Ventilation and air quality
Good Practice: ▸ D3.9.1 ▸ D3.9.2 ▸ D3.9.3 ▸ D3.9.4 ▸ D3.9.5 ▸ D3.9.6	*To ensure adequate ventilation and air quality, take the following actions in a coordinated manner.* • Design to achieve the ventilation rates recommended by the CIBSE Guide (1987) Section B2: Ventilation and Air Conditioning requirements. • Review D3.9.1 for Air Infiltration and Ventilation Rates. Incorporate appropriate measures. • Review D3.9.2 for Natural Ventilation and Passive Stack Ventilation in Small Buildings. Incorporate appropriate measures. • Review D3.9.3 for Ventilation in Large Buildings and incorporate appropriate measures. • Review D3.9.4 and D3.9.5 for Mechanical Ventilation and Heat Recovery and Air Conditioning and incorporate appropriate measures. • Build in ease of maintenance and inform the client of his obligations in this respect. Poor maintenance has been identified as the root cause of many of the environmental problems in buildings (Building Use Studies Survey of 47 Buildings in the UK). Maintenance must be carried out promptly and regularly to avoid health problems in any building, but particularly those with air conditioning. • Select building equipment for ease of maintenance. Preventative maintenance regimes must be evolved by the designers and communicated to the users. • With increased air-tightness of building envelopes, controlled ventilation (whether natural or mechanical) is essential. The use of trickle ventilators incorporated into the window frames may be required and can help to combat condensation. • In buildings (except industrial), tobacco smoke is often the most obvious indoor air contaminants. In commercial and public buildings, exposure can be limited by providing special smoking areas, or banning smoking from the workplace. • Although there is much documentary guidance on indoor air quality and ventilation, there is little consistency in this advice, and standards are continually under review. Designers and specifiers must seek out the latest available guidance applicable to their project. • It may be possible to reduce the quantity of outdoor air drawn into the building via a central air handling plant (if required) by providing a filtration system. In sealed buildings occupied by non-smoking people this will entail removing bio-effluents and CO_2 expired by the occupants and the normally low concentrations of other contaminants released by the fabric, furnishings and cleaning operations. Filtration load can be reduced by locating the fresh air inlet as far as possible from sources of contamination, such as local traffic.
Good practice references and further reading:	• See background references, plus: • As listed in D3.9.1 to D3.9.6. • CIBSE Guide, Section B2, 1987. • ASHRAE Standard 62–1989: *Ventilation for Acceptable Indoor Air Quality* (American Society of Heating Refrigeration and Air Conditioning Engineers).

Issue D4.6.5	Lighting
Background: ▸ D3.10 ▸ D4.6.9	Refer to D3.10 for general background. As can be seen from this, designing to maximise daylighting is desirable both in terms of occupant preference and also to reduce energy consumption. The Energy Efficiency Office estimate that lighting costs in the commercial sector are generally higher than heating costs, though they can be reduced more easily. Large reductions in lighting energy are possible by increased use of daylight, together with improved control of artificial lighting systems (which can take many forms). In turn, estimated CO_2 emissions by end-use (Friends of the Earth 1987) from the energy used for lighting is 6.5%. This is excluding any emissions from manufacture, transportation and installation, and indicates that this element is a significant contributor to the greenhouse effect. However, increasing daylighting can have several unwanted consequences if not considered carefully: • increased heat loss through glazing (Refer to D4.3.3 for relative U-values). Use low emissivity double or triple glazing to reduce this loss; • heat gain in the summer, particularly on south or south west elevations. This can result in the need for mechanical ventilation or even air conditioning in extreme cases; • excessive brightness (glare). In addition, poor lighting design is thought to be a contributing factor to so called Sick Building Syndrome. See D4.6.9 for details on this. The provision of good task lighting at workplaces helps reduce the need for high overall light levels, and allows occupants some control over their visual environment. New energy efficient lamps, good reflectors and diffusers, and good controls with timed operation can often show major savings on energy costs, particularly for those lights which are switched on for the longest periods.
Background references:	• Curwell, S.R., March, C.G. and Venables, R.K., (Eds), *Buildings and Health, The Rosehaugh Guide to the Design, Construction, Use and Management of Buildings*, RIBA Publications, 1990. • Department of Environment Energy Efficiency Office: Best Practice Programme: Guides and Case Studies.

Issue D4.6.5	Lighting
Good Practice: ▸ D3.10 ▸ D4.6.9	• Review D3.10 and incorporate any relevant measures. In particular, ensure the design achieves good daylighting indoors. • In commercial buildings, aim for a ratio of about 20% of the wall area to be glazed, with double-glazed thermally-broken frames. (To reduce heat loss further, use low emissivity glass. Check the Building Regulations, 1991, Approved Document L for areas of glazing). • Specify activity sensors so that lights are turned off automatically when activity stops, leaving them to be turned on again manually. • Specify good quality, efficient, high-frequency ballast light fittings and luminaires with appropriate diffusers in preference to conventional switch start lights. (Refer to D4.6.9 for references to lighting and Sick Building Syndrome). • Specify good task lighting at workplaces to reduce the need for high overall light levels. Select appropriate lighting levels for the activity. This saves energy and gives visual contrast to the spaces as well as giving occupants as much control as possible. • Provide for local switching to allow occupants to turn off lighting in areas where it is not required. • Specify compact fluorescent lamps and aim for a variety of fittings and variety of colour rendering, using low energy downlighters etc. (Compact fluorescent fittings can give a light more like that of incandescent lamps but with an 80% energy saving, so reducing energy consumption and internal heat gains). • Design for solar controlled external sun blinds as appropriate, to reduce heat gain and/or glare. • Consider deflector silvered blinds internally in high level glazing to reflect daylight as far into the building as possible. Or alternatively, use an external 'light-shelf' to reflect light onto the ceiling. • Consider installation of timers to automatically switch off lighting at lunchtime, and in the evening and night if appropriate (i.e. in commercial buildings). • Reduce external lighting to minimum levels suitable for safety and security without being wasteful of energy. Avoid high levels of floodlighting. • Advise on the management policy associated with the design and use of controls, in terms of energy use. • Choose internal decorations carefully to suit the lighting design (Refer to D4.6.9 for further information).
Good practice references and further reading:	• See background references, plus: • *CIBSE Code for Interior Lighting*, CIBSE, 1984. • *CIBSE Lighting Guides* covering The Industrial Environment, Hospitals and Health Care Buildings, Areas for Visual Display Terminals, Sports, Lecture Teaching and Conference Rooms, and the Outdoor Environment, Chartered Institution of Building Services Engineers, in sections, 1989–1991. • *Energy Efficient Lighting for Buildings*, BRECSU–THERMIe, 1992. • Littlefair, P.J., *Site Layout planning for daylight*, BRE IP 5/92. • Littlefair, P.J., *Site Layout for sunlight and solar gain*, BRE IP 4/92.

Issue D4.6.6	Condensation
Background:	Changes in temperature or moisture content can cause condensation to occur. This can happen when warm damp weather follows a period of cold, but is more usually caused by people and activities inside a building. Moisture vapour released by cooking and clothes washing in dwellings can condense onto cold surfaces. Many industrial processes also require high humidities and temperatures and release large quantities of steam.
	Condensation happens in two ways. First, when warm moist air comes into contact with cold surfaces which are below its dew-point, water will condense upon them. Secondly, when the material of a wall, roof or other building element is permeable to water vapour, which applies to most building materials, then interstitial condensation can occur if at any point within the construction, the temperature falls below the dew-point for the water vapour present in it. Under-heating coupled with minimal ventilation is the major cause of the problem. Most often a problem in dwellings, surface condensation encourages moulds and fungal growth as well as unpleasant smells. Warm humid conditions are also ideal breeding grounds for bacteria. Interstitial condensation can lead to saturation of the building fabric with consequent deterioration, loss of thermal performance (which exacerbates the problems of internal surface condensation) and structural damage. Fabric degradation caused by interstitial condensation is of particular importance in the following situations and an assessment of the risk of such condensation must be made: • where structural components are liable to rot, e.g. timber framed walls or roofs, or are liable to corrode, e.g. steel framed walls; • in constructions with impermeable outer layers, e.g. walls with impermeable cladding; • in constructions containing metal reinforcement, e.g. reinforced concrete; • in buildings such as swimming pools likely to have high internal vapour pressures.
Background references:	• BRE Digests: No.110, *Condensation*, 1972 Edition; No.270, *Condensation in insulated domestic roofs*, 1983; No.297, *Surface condensation and mould growth in traditionally built dwellings*, 1985; No.336, *Swimming pool roofs: minimising the risk of condensation using warm deck roofing*, 1988; No.369, *Interstitial condensation and fabric degradation*, 1992.
Good Practice:	*To reduce the risk of surface condensation occurring, consider the following when carrying out detail design:* • adequate provision for ventilation, by the use of extractor fans or passive stack ventilation in kitchens and bathrooms, and controllable natural ventilation; • reduce internal water vapour content in high risk buildings, such as swimming pools, by specifying forced ventilation or dehumidifiers; • good thermal insulation to bring internal surface temperatures close to air temperature, thereby reducing the risk of surfaces falling below dew-point; • provision of central heating in dwellings; • use materials on walls/ceilings that permit moisture to disperse and surfaces to dry. *In addition to reduce the risk of interstitial condensation:* • assess the risk using a micro-computer software package, such as BRECON II; • include effective vapour barriers on the warm side of the structure where necessary; • allow for cavity ventilation in cold deck flat roofs and pitched roofs, and on drylined solid walls; • specify warm deck roofs in high risk buildings such as swimming pools; • ventilate behind external claddings and provide drainage channels to allow condensate to escape; • specify an effective vapour barrier on the warm side of timber framed walls, and consider including breather paper to protect sheathing from external moisture ingress if a lightweight cladding is used externally; • when refurbishing existing buildings, assess the risk of interstitial condensation from any proposed remedial measures.
Good practice references and further reading:	• See background references, plus: • Chartered Institution of Building Services Engineers, *CIBSE Guide* Section A10. • *Interstitial Condensation and Fabric Degradation*, BRE Digest 369, 1992. • BS5250: 1989: *Code of Practice for control of condensation in buildings*. • BS6229: 1982: *Code of Practice for flat roofs with continually supported coverings*.

Issue D4.6.7	**Controls**
Background:	The selection of appropriate controls for heating, ventilation, air conditioning (HVAC) and lighting is essential to help reduce energy use, and to provide a comfortable and responsive environment within a building.

A heating and hot water control strategy should be determined which takes account of the pattern of use of the building, and allows occupants to obtain the comfort conditions they need. It should also allow plant to operate at high efficiency throughout its working range and distribute heat around the building with a minimum of waste. Control system design should ensure that central plant is off when not required, with zoning where necessary.

Modern electronic controls and Building Energy Management Systems (BEMS) can be comprehensive and precise in controlling HVAC systems. However, their performance must not be taken for granted and installations must be carefully specified, installed, commissioned and maintained. Their parametric requirements and limits also need to be carefully defined.

Lighting controls should also be considered, for instance to switch lights off when activity stops, to suit patterns of use within a building or to only allow them to be switched on which daylight is insufficient. In some areas, lighting can be time-controlled during peak hours and at other times respond to movement and other types of sensor.

It is important that occupants and users understand controls and how best to use them. Clients, for instance, should understand that building staff must be sufficiently trained and skilled to make full use of the capabilities of BEMS. Individual control over local temperature, ventilation rates and lighting levels is also vital to avoid occupant discomfort and their purpose must be explained to the occupants. |
Background references:	• *Energy Efficiency in Offices: Good practice case studies 13, 14, 20, 21,* Energy Efficiency Office Best Practice Programme, Department of the Environment, May and June 1991.
Good Practice:	*When specifying controls consider the following:* • provision of a detailed controls specification; • matching controls to the intended usage or tenancy patterns; • keeping manual controls simple and user friendly and put them in sensible places; • designing heating, ventilation and air conditioning systems so that areas with different usage requirements or different patterns of solar gain (for example on opposite sides of a building) are zoned separately; • fitting thermostatic radiator valves wherever possible; • provision of on-demand controls in intermittently-used areas, with automatic switching-off when activity ceases; • provision of controls which permit energy monitoring and target setting in buildings; • using of electronic BEMS controls for heating, ventilation and air conditioning systems in public and commercial sector buildings; • including a requirement for mechanical and electrical service contractors to provide full details of the control systems and their operation for inclusion in a building user or occupant manual, and ensure they are provided before the building is handed over to the client.
Good practice references and further reading:	• BEMS Centre, *Standard Specification,* Volumes 1 and 2, BSRIA. • BEMS Centre, *Code of Practice for the commissioning of Building Energy Management Systems,* BSRIA. • *Automatic controls and their implications for system design,* CIBSE Application Manual AM1, Chartered Institution of Building Services Engineers, 1985. • Good Practice Guide 46, *Energy Efficiency in Offices,* EEO, November 1992. • Halliday, S.P., *Environmental Code of Practice for Buildings and Their Services,* BSRIA, 1994. • *The BEMS Book,* BSRIA. • Honeywell, *Engineering Manual of Automatic Control,* 1989. • Levermore, G.J., *Building Energy Management Systems – An application to heating and control,* E & F N Spon, 1992.

Issue D4.6.8	Legionnaires' Disease

Background:	This is a type of pneumonia – a rare human disease which has only been recognised since the late 1970s. It derives from inhaling *Legionella* bacteria which are present (in small quantities) throughout the world wherever there is natural fresh water. Normally, the bacteria present no risk to humans, but in the last 10 to 15 years, it has become apparent that the bacteria when entering building water systems, may colonise them and grow to high concentrations. The bacteria can be inhaled if the water is sprayed into the air, forming an 'aerosol' of bacteria, which is exactly what happens in the normal operation of evaporative cooling towers, and in the operation of showers. It can also occur when running hot and cold water at normal pressure into baths and basins, particularly with spray taps.
	The infection is more likely to develop in the old and infirm than the young and healthy, and occurs more often in men than women. Initially it was observed in older people, particularly hospital patients, but it is now known to occur in the more general population. Fortunately, the number of cases per year is small (about 2% of all community-acquired pneumonia in Britain), although as many as 1 in 10 may die of the illness unless it is recognised and properly treated.
	For the designer, it is important to appreciate that any deaths or illnesses from Legionnaires' Disease, which may result from poor building or plant design and/or maintenance, are preventable. The conditions allowing the growth of *Legionellae* bacteria are:
	• water temperature in the range of 20–50°C (optimum being between 30 and 40°C); • stagnation or a long stay of water in a system (this need not be within the whole system, but may just be in dead-legs or little-used pipework branches); • food sources within the system, either as inappropriate materials incorporated in the construction of the system or as contamination from environmental sources into stored water.
	The control of *Legionella* is now well understood, and is essentially a function of the management of good microbiological water quality. Guidance is published in the UK by the DHSS (1988), the HSC (1991) and CIBSE (1989 and 1992).
	Evaporative Cooling Systems for air conditioning (i.e. wet systems) have been the source of many outbreaks of Legionnaires' Disease although showers and spray taps cause more outbreaks than air conditioning systems. This is often resulting from supplies or systems being reactivated after having been idle for some time. Water in these systems recirculates and is continually contaminated with material washed from the air, drawn into the cooling tower, (such as dust, fungi, bacteria, insects, leaves and pollen). This is added to by a small but constant supply of micro-organisms, delivered by the water supply. As water is evaporated from the tower, the contaminants are left behind in the circulating water, where they accumulate. This can only be controlled by starting with good quality softened water, by bleeding some of the water to the drain and replenishing it with good quality top-up water, and by ensuring a properly managed water treatment regime is followed.
	The same contamination, i.e. from the air, and from supply water can lead to microbiological growth within tanks in cold water storage installations. By maintaining cold water storage below 20°C and by excluding external contaminants there will be little opportunity for micro-organisms to accumulate.
	In hot water cylinders, the main problem is accumulation of debris at the bottom of the cylinder. The temperatures in the cooler areas below the heating element may encourage microbiological growth. Selection of appropriate plant without these dead areas, together with regular flushing through is essential to avoid this growth.
	Ventilation systems also need to be considered carefully due to their use of water in two ways – humidification, and condensation. Although generally not implicated as sources of Legionnaires' Disease, these systems are frequently mis-managed and may be sources of both microbiological and chemical pollutants in indoor air.
	Spa baths or jacuzzis have been identified as sources of legionella infection in the past. This is due to the recirculation of warm water which can lead to a build up of micro-organisms. They must be rigorously maintained, by draining and cleaning daily, and the strainer removed and cleaned on every occasion. A suitable chlorine level must always be maintained in the water.
Background references:	• As good practice references, plus: • Brundritt, G.W., *Legionella and building services*, Butterworth-Heinemann, 1992.

Issue D4.6.8	Legionnaires' Disease

Good Practice:	*To keep the potential for Legionnaire's Disease to a minimum by good design, when specifying services using water:*

- maintain water temperatures either below 20°C, or above 55°C in stored water;
- ensure stored water is insulated against heating from the building, particularly if sited near boiler rooms, although smaller domestic tanks of less than 500 litres need not be insulated underneath due to the fast turnover of water;
- maintain water temperature at outlets at 50°C;
- avoid intermediate temperatures in stored water;
- exclude external environmental contaminants to maintain good quality stored water;
- with wet cooling systems, bleed the water periodically and top-up with good quality water;
- follow maintenance recommendations thoroughly;
- exclude from water systems materials of construction which act as food sources;
- install a fitted lid to cold water tanks to exclude external contaminants, and keep out light (which algae etc rely on for growth);
- construct the system of approved materials (consult the Water Byelaws Advisory Service: *Water Fittings and Materials Directory*, 1993);
- match capacity to demand to ensure the highest possible through-put of water, ideally with a turnover time of no more than 8 hours;
- set the maximum residence time at 24 hours;
- disconnect outlets used less than once in 60 days from the system at the branch in accordance with BS6700;
- estimate the water demand of a system approximately by calculating a maximum daily usage for each outlet and totalling the results and then adjust the capacity of the cold water storage and of the hot water cylinders to the maximum calculated daily demand so that the equivalent throughput of only one day's supply of water is kept in the building;
- design pipework layouts so as to eliminate stagnation or 'dead-legs';
- ensure that air intake for a ventilation system is remote from the cooling towers to avoid possible cross-contamination in wet cooling systems;
- flush through hot water cylinders regularly using a low level feed and drain point to clean away accumulated debris;
- ensure efficient drainage of driptrays from ventilation systems, and avoid any stagnant water;
- provide easy access and lighting for frequent inspection and cleaning;
- drain and clean systems as required by the manufacturer, including cleaning filters and strainers etc. carefully;
- maintain chemical treatment (if appropriate) to designed levels e.g. residual chlorine level in jacuzzis, although routine use of chemicals in water distribution systems is not recommended;
- water softeners must be back-flushed (purged) once every 24 hours and disinfected once every six months;
- drain any water systems left unused for many days and leave them empty or at least drained and refilled before being used again, an issue especially important in schools, colleges and halls of residence where, in most cases, chemical cleaning may be required before refilling.

Good practice references and further reading:	

- BS6700: 1987 *Specification for Design, Installation, Testing and Maintenance of Services Supplying Water for Domestic Use Within Buildings and their Curtilages*, BSI, Milton Keynes.
- Curwell, S.R., March, C.G. and Venables, R.K., (Eds), *Buildings and Health, The Rosehaugh Guide to the Design, Construction, Use and Management of Buildings*, RIBA Publications, 1990. Chapter C3.
- Chartered Institute of Building Services Engineers, *Minimising the Risk of Legionnaires' Disease*, Technical Memo 13, 1987.
- *Water Fittings and Materials Directory* 1993, Water Byelaws Advisory Service.
- *Microorganisms in Building Services*, BSRIA TN3, 1988.
- *Health and Hygienic Humidification*, BSRIA TN13, 1986.
- *The control of legionellosis including legionnaires' disease*, HSE, 1991.
- Stewart, H., *Air Conditioning Systems and Legionnaires' Disease*, Conference Studies, London, June 1989.
- Hejab, M. and Parsloe, C., *Space and weight allowances for building services plant*, TN9/92, BSRIA, 1993.

Issue D4.6.9	**Sick Building Syndrome**
Background:	Some people at work are known to suffer a sickness or malaise which seems to have no specific cause, but which appears to be related to the building in which they work. This has become known as sick building syndrome (SBS). Although it can affect a large number of occupants, it is difficult to quantify the extent of its overall affect. It can cause loss of productivity, perhaps by as much as 20%, and it can affect over half the staff in a 'sick building'. Although the precise causes of SBS are unknown, it appears to be caused by a number of factors which act cumulatively on individual office staff and result in a typical set of symptoms. The main categories of possible contributing factors are: *Poor design:* • low floor-to-ceiling height; • large unstructured open plan offices; • absence of natural light; • poor lighting scheme. *Poor air and environmental quality:* • poor air quality; • poor lighting; • overheating or underheating; • noise. *No personal control over work or environment:* • stress; • sedentary and repetitive work; • HVAC systems; • lighting. *Poor management and maintenance:* • poor cleaning, adjustment, operation. In addition to the above, computer use and passive smoking may be contributory factors. The relationship between computer use and building sickness is not strong, although those who worked for more than six hours per day at a VDU may have a slightly higher sickness rate. However, staff subjected to the effects of other people's smoke – so-called passive smoking – appear to have higher sickness rates than those not exposed to this. SBS poses no threat to construction site workers or to members of the public who pass by, or even visit (short-term) affected buildings. It is experienced solely by building occupants, and of these, it is female clerical staff doing routine sedentary work who are most affected. When people are doing varied work, or are moving around (even within the 'sick building') they are less likely to be affected. But, if conditions are particularly poor, staff at all levels can be affected. The characteristic symptoms fall into five categories: • irritation of eyes, nose and throat; • irritation of the skin; • neurotoxic effects: mental fatigue, headaches, lethargy, tiredness; • respiratory effects: asthmatic symptoms; • effects on senses of smell and taste. These symptoms occur when people are at their place of work, but disappear in the evening and at weekends. It is interesting to note that experience of the symptoms may not be continuous, and that it is persistent and chronic rather than acute. It is more likely to occur later in the working week, and there is no evidence to suggest any seasonal variation in the syndrome, or that it causes any long-term health effects.
Background references:	• Curwell, S.R., March, C.G. and Venables, R.K., (Eds), *Buildings and Health, The Rosehaugh Guide*, RIBA Publications, 1990., Chapter C5. • Finnegan, M., J., Pickering, C.A.C., and Burge, P.S., The Sick Building Prevalence Studies, *British Medical Journal* 289, 1573–5, 1984.

Good practice references and further reading:	• Potter, I.N., *Sick Building Syndrome*, Report BTN04/88, BSRIA, 1988.
	• *Buildings and Health* as above.
	• *Sick building syndrome: a review of the evidence on causes and solutions*, HSE Contract Research Report 42/49, HMSO, 1992.
	• Holdsworth, W., Sealey, A.F., *Healthy Buildings – A design primer for a living environment*, Longman, 1992.
	• CIBSE (Chartered Institution of Building Services Engineers):
	  (a) *Environmental Criteria for Design*, Guide Section A1.
	  (b) *Ventilation and Air Conditioning Requirements*, Guide Section B2.

Issue D4.6.9	**Sick Building Syndrome**

Good Practice:	SBS is best controlled by exercising high standards of design, cleanliness, maintenance and management. The following guidance deals with the four categories of possible contributing factors in turn.

Design

▶ D4.3.2
▶ D4.6.4

- Avoid air conditioning and mechanical ventilation systems if possible (refer to D4.3.2 and D4.6.4) but, if required, ensure that adequate outdoor air is supplied. Ensure that air inlets bringing primary air into a building are clear of any sources of pollution and contamination. Then ensure that adequate air is supplied to all parts of the building.
- Ensure the outlets or extracts are correctly positioned to avoid air short-circuits in and out of rooms without reaching the occupied zones.
- Ensure adequate exhaust and cooling capacity from the occupied areas.
- Ensure good access is available to plant rooms for maintenance.
- Regarding layout, occurrences of SBS symptoms are greater in shared deep plan offices (with work spaces say more than 15 m from a window) and with low floor to ceiling height (less than 3 m). Where possible avoid designs outside these limits.
- Regarding interior design, lighting is very important. Lighting schemes should not be monotonous but provide uplift for visual interest. Avoid intense colours which are visually very demanding. Glare and flicker result in eye strain and headaches, whereas severe deficiencies in lighting will cause stress. Use high frequency fluorescent lights to avoid flicker, with high quality diffusers to avoid glare. Daylight is always preferable, but avoid too much sunlight because it causes overheating and glare problems. Provide local task lighting for user control.
- Provide areas of visual relief and interest, e.g. lightwells, atria, fountains etc.
- Avoid use of tinted or reflective glazing, if possible, as these reduce psychological links with external conditions.
- Isolate and separately ventilate office machinery – photocopiers, laser printers, plotters etc when they are used in clusters or continuously during the day.

Air and environmental quality

- If air conditioning is essential, avoid the following:
  - inadequate or dirty air handling ducts and filters;
  - malfunctioning humidifiers or disconnected damper linkages;
  - bad re-adjustment when loads change (e.g. after layout alterations);
  - over-complex control systems, or over-ridden automatic control systems;
  - excessive noise.
- Provide adequate resources for commissioning, operation and maintenance (including allowance for re-commissioning occasionally).
- Ventilation rates given by the Chartered Institution of Building Services Engineers should be used as a minimum standard.

Personal control over environment

- Provide as much individual control over the work place conditions as possible.

Management and maintenance of systems:

- Operate the environmental control systems efficiently.
- Maintain the environmental control system regularly (see point above for maintenance of air conditioning).
- Ensure an efficient system exists for reporting environmental problems and follow up remedial action promptly.
- Prioritise environmental standards, placing highest emphasis on areas occupied by those doing sedentary work.
- Periodically monitor environmental conditions.
- Introduce smoking controls in working areas.
- Regularly 'deep-clean' carpets and fabrics which may trap particulates, dirt, and dust mites.

D4.7 Materials

Issue D4.7.1	Use of environmentally acceptable materials
Background: ▸ D3.8.2 ▸ D4.7.5	See D3.8.2 for general background to the green labelling of materials and products. Developing an eco-label for comparison of similar materials is not straightforward. In general, there are few common scales for comparison of environmental impacts. 'Eco-labelling is inevitably based on contestable value judgements about the relative significance of difficult-to-compare environmental impacts'. (Evans, 1993). Careful specification of timber from sustainable resources is an area where designers can make a significant contribution to using an environmentally acceptable material without de-forestation. Studies by the International Tropical Timber Organisation in 1990 indicated that less than 0.2% of tropical rain forests are being managed sustainably for commercial production. Refer to D4.7.5 for more information on this subject.
Background references:	• Evans, B., Rating Environmental Impact, article in *Architect's Journal*, 19 May 1993, p.46. • Atkinson, C.J., and Butlin, R.N, *Ecolabelling of building materials and building products*, BRE Information Paper 11/93, BRE, 1993. • Construction Industry Environmental Forum, *Materials and product blacklists*, Notes of a meeting held on 23/7/92, CIRIA, 1992. • Construction Industry Environmental Forum, *Purchasing and specifying timber*, Notes of meeting held on 23/02/93, CIRIA, 1993. • Construction Industry Environmental Forum, *The use of timber in construction*, Notes of a meeting held on 23/9/93, CIRIA, 1993.
Good Practice: ▸ D3.8.2 ▸ D4.7.5	*At this detailed design stage:* • remind yourself of the importance and problems in this area by reviewing D3.8.2; • follow the Good Practice guidelines in D3.8.2; • review the EC Directives on Dangerous Substances and the EC Regulation on Ecolabelling and confirm project policies on selection of materials; • in the absence of an environmental labelling scheme, seek appropriate information from manufacturers when making primary choices about materials; • when specifying timber, refer to the notes of the Forum meeting on purchasing and specifying timber and the good practice references listed and consider making greater use of timber from well-managed forests; • review D4.7.5 for more information on sourcing of timber.
Good practice references and further reading:	• Fox, A. and Murrell, R., *Green Design: A guide to the environmental impact of building materials*, Architecture Design and Technology Press, 1989. • *Timber: Eco-labelling and Certification*, Publication L295, Friends of the Earth, 1993. • *Timber: Types and sources*, Publication L296, Friends of the Earth, 1993. • *The Good Wood Manual*: Specifying Alternatives to Non-renewable Tropical Hardwoods, Friends of the Earth, January 1990. • Hall, K and Warm, P., *Greener Building Products and Services Directory*, Association for Environment Conscious Building Directory, Second Edition, 1993. • Curwell, S.R. and March, C.G. (Eds), *Hazardous Building Materials – A guide to the selection of alternatives*, E & F N Spon, 1986. • UK Ecolabelling Board – Tel: 071-820 1199. • Forthcoming report (1994) from CIRIA Project 461: Environmental criteria for selection of construction materials.
Legal reference:	• EC Regulation (880/92) on a Community Eco-Label Award Scheme.

Issue D4.7.2	Environmental policy on materials
Background:	Increasing concern is being shown regarding the environmental impact of building materials. The impact on people, particularly in terms of health and safety, throughout a particular product's life-cycle is a key issue. Asbestos is an obvious example but concern is also being raised about man-made mineral fibres, and dust and vapours from other products. Other environmental concerns regarding materials include: • **embodied energy** – including energy used in manufacture and transportation; • **sustainability** – if a natural resource is being mined or quarried, is it being used in a way which reflects and respects its finite availability? – this includes trees such as tropical hardwoods which in some parts of the world are being harvested in an unsustainable way; • **effect on landscape, wildlife and/or local communities** of quarrying and other mineral extraction; • **use of materials inappropriate to the location;** • **the international and global environmental implications** of using certain materials, for example CFCs. The environmental policy for the project should cover such issues as they relate to construction materials. It may even provide explicitly for the use or non-use of defined materials.
Background references:	See good practice references.
Good Practice:	*At this stage, consider the following actions:* • review the environmental policies for the project, and indeed those for the organisations involved in the project, for specific materials-related provisions and take appropriate action; • closely examine the environmental information and records of likely materials suppliers; • the possible development of a project database of environmentally acceptable materials and/or suppliers.
Good practice references and further reading:	• Curwell, S.R., March, C.G. and Venables, R.K., (Eds), *Buildings and Health, The Rosehaugh Guide to the Design, Construction, Use and Management of Buildings*, RIBA Publications, 1990. • Curwell, S.R., Fox, R.C. and March, C.G., *Use of CFCs in Buildings*, Fernsheer Ltd, 1988 (out of print). • Curwell, S.R. and March, C.G. (Eds), *Hazardous Building Materials – A guide to the selection of alternatives*, E & F N Spon, 1986.

Issue D4.7.3	Criteria for selection

Background:	A primary concern in selecting materials for effective, environmentally sensitive use in construction is to consider the life cycle environmental costs of the material and its use. Tools to make such comparisons are limited at present but under active development, including a study by CIRIA under the auspices of the Construction Industry Environmental Forum on environmental criteria for selection of construction materials, in which outline life-cycle environmental impact analyses of the most-used construction materials are being developed.
	Against the background of reducing energy consumption and pollution, another key issue here is to assess the relative significance of energy consumed in the manufacture of buildings compared with that used directly in heating and lighting etc. There has been little recent research into this problem, most data being based on sources in the 1970s. BRE and Davis Langdon and Everest are currently undertaking further analysis on this using the latest data, but their results are not yet publicly available.
	The energy issues involved are complex, and broadly as follows:
	• energy used in winning raw materials; • embodied energy in manufacture and/or processing; • pollutants (during manufacture, and in some cases thereafter); • transport; • re-use;
	and these factors form a very major part of the life cycle analysis already mentioned.
	Current data available suggests that the total energy required to manufacture the building materials for, and to construct, a 3-bedroom semi-detached house of 100 m² area of masonry construction is equivalent to the energy consumed over 3 years for heating and lighting the same property. If we assume a life expectancy of the building of 60 years, then the 'construction' energy needs are shown to be only one-twentieth of the total energy requirements of the building throughout its life. It has also been calculated that conservation of energy resulting from higher levels of insulation has a greater effect on the total energy consumption, compared with that used in making the materials. So one conclusion would be to concentrate effort into reducing direct energy consumption. But, as the energy efficiency of buildings is improved, the indirect energy consumption becomes a bigger proportion of the total. Combine this with a likely reduction in the life span of buildings (particularly industrial and commercial) and the effect is to increase the significance of the indirect component. So, as the construction industry moves towards well insulated energy efficient design standards, it becomes increasingly worthwhile for designers to consider the energy consumed in the manufacture of the buildings.
	A large number of environmentally objectionable materials used in the building industry have been withdrawn from production, or modified in ways which, quite incidentally, makes them more environment-friendly because of the need to meet more stringent health and safety regulations and standards during their manufacture. These regulations are under continuous review and the benefits to the environment which result will be quite considerable.
	At present it is still a material's marketability which drives its production or extraction, rather than a view of the balance between the desirability of the end product, and the consequences of extracting the raw material.
Background reference:	• Evans, B., Rating environmental impact, in *Architects Journal*, 19 May 93, p46.

Issue D4.7.3	Criteria for selection

Good Practice:	Lack of up-to-date data, combined with the complexities of the analysis, make realistic, energy or emissions-based selections in routine design impossible at present. A clear system that could be applied in design in order to carry out full energy audits of both direct and indirect energy consumption will become increasingly important in the future.
	What can be achieved now is to act on following points.
	• For relative energy costs between materials and components, refer to Table D3.4 in *Buildings and Health*, p.448 – although based on out-of-date data, the comparative figures may still be relevant.
	• Contact BRECSU for advice and data on embodied energies of construction materials.
▸ D4.7.5	• For process energy data for construction materials, limited to wood, steel, concrete and aluminium, see Table 11, in Tropical Hardwoods, Fruits of the Forest, article in *Architects' Journal* 8 August 1990, p.52.
▸ D4.3.3	• If timber can be obtained from sustainably-managed sources, its use can be promoted as a material having a low environmental impact – see D4.7.5 for details on this issue.
	• Seek to select materials from a local source to reduce transport costs, especially if it is otherwise necessary to import an alternative from another country.
	• Select appropriate insulation materials (see D4.3.3 for guidance on indirect energy costs).
	• Seek to select materials which are as natural as possible, i.e. those that have not changed much during processing.
	• Select renewable or reusable raw materials where possible.
	• Where the data is available, use materials that have minimal impact on the environment during mining or extraction.
	• Because the potential lifespan of a structure or building is crucial, consider designing for flexibility to allow the life-span to be extended so clearly reducing the demand for new materials – 'designed obsolescence' or 'built-in obsolescence' are notions opposed to caring for the environment.
	• If high-energy materials, such as metals, are to be used, employ them in ways which facilitate recovery in due course and, in particular, try to avoid using aluminium made from fossil-fuel-generated electricity (unless it is recycled) if an acceptable alternative can be used.
	• Refer to *Green Design* by Avril Fox and Robin Murrell – although exact comparisons cannot be made at present, guidance on 'low energy use', 'inherently durable' and 'limited potential for recycling' is given for general comparisons.
	• Use the *Greener Building Products and Services Directory* and any other appropriate sources of information on 'green' construction products.
▸ D4.2.2	• Avoid hazardous building materials; refer to D4.2.2.

Good practice references and further reading:	• Hall, K and Warm, P., *Greener Building Products and Services Directory*, Association for Environment Conscious Building Directory, Second Edition, 1993.
	• Fox, A. and Murrell, R., *Green Design: A guide to the environmental impact of building materials*, Architecture Design and Technology Press, 1989.
	• Ove Arup & Partners, *The Green Construction Handbook – A Manual for Clients and Construction Professionals*, JT Design Build, Bristol, 1993.
	• Curwell, S.R., March, C.G. and Venables, R.K., (Eds), *Buildings and Health, The Rosehaugh Guide to the Design, Construction, Use and Management of Buildings*, RIBA Publications, 1990.
	• Tropical Hardwoods, Fruits of the Forest, article in *Architects' Journal*, 8 August 1990.

Issue D4.7.4	Avoidance of CFCs, HCFCs and reduction in NO_x and CO_2 and SO_x
Background:	*Chloroflurocarbons (CFCs)* – These gases have become widely known in the latter part of the 20th century as causing damage to the ozone layer in the Earth's upper atmosphere and act as so-called greenhouse gases. First used in the 1930s as refrigerants, CFCs have been used widely in aerosol propellants, foaming (or blowing) agents for plastics and insulants, in air conditioning plant, refrigerators and fire extinguishers. The effect of CFCs on the ozone layer was predicted in the early 1970s, but not first observed as a 'hole' over Antarctica until the late 1980s. The effects are to allow more of the sun's damaging ultra-violet radiation through to the Earth's surface, which has and will further increase cases of skin cancer. About 50% of the world's production of CFCs is used in buildings and so designers and specifiers can have a large influence over their reduction. An international agreement of 1987 entitled The Montreal Protocol and related EC Regulations were revised in late 1992 to bring forward the CFC phase-out dates. Within the EC, the production and import of CFCs are to end by 1 January 1995. After this date, the only source of CFC refrigerants will be from stockpiles and CFCs recovered from redundant equipment. BRE Digest 358 (February 1992) warned that '... some industry sources predict a shortage of some CFC refrigerants much earlier. Owners and operators of CFC equipment are advised to plan urgently for CFC phase-out.' *Hydrochloroflurocarbons (HCFCs)* – These gases contain chlorine, but have low ozone depletion potentials (ODP). They have a shorter atmospheric lifetime than CFCs (because they contain hydrogen) so are thought to be much less damaging to stratospheric ozone. These are also subject to controls under the Montreal Protocol and EC Regulation, which sets a cap on their consumption in 1996 and a gradual reduction to 10% of that by 2015. However, this may be accelerated by the EC, as was the Montreal Protocol Schedule for CFCs. *Nitrogen Oxides (NO_x)* – These are pollutants put into the atmosphere by fossil fuel combustion, in particular from vehicle exhausts. Its effect is to form diluted nitric acid which can kill trees and other plants. A large proportion of Germany's forests are dead or dying because of acid rain. *Sulphur Dioxide (SO_x)* – Another pollutant derived from burning fossil fuels, this one from coal-fired power stations. This again, like nitrous oxides, forms acid rain, further reducing forests. *Carbon Dioxide (CO_2)* – The most important greenhouse gas, responsible for the warming of the Earth's atmosphere. CO_2 enters the atmosphere when fuels containing carbon are burned. Coal, wood, natural gas and oil all contain carbon and all produce CO_2 in use. Burning of rainforests is another major sources of CO_2 emission resulting from land clearance for ranching or agriculture although not linked directly with energy consumption.
Background references:	• Good practice references, plus: • Department of the Environment, Pollution Paper No.5, *Chloroflurocarbons and their Effect on Stratospheric Ozone*, HMSO, 1976. • Curwell, S.R., Fox, R.C. and March, C.G., *Use of CFCs in Buildings*, Fernsheer Ltd, 1988 (out of print). • *The Montreal Protocol on Substances that Deplete the Ozone Layer*, Montreal 16/09/87, Foreign Office Command Paper, Treaty Series No.19, HMSO, 1990, and subsequent amendments.

Issue D4.7.4	**Avoidance of CFCs, HCFCs and reduction in NO$_x$ and CO$_2$ and SO$_x$**

Good Practice:	*The simple answer is to reduce energy consumption. Any of the following measures towards this objective will help:*
	• avoid products using CFCs in their manufacture and generally use environmentally acceptable materials;
	• avoid air conditioning if at all possible, particularly if it uses fully halogenated CFCs;
	• use alternatives to CFCs, such as HCFCs or HFCs if converting an existing CFC-based air conditioning system;
▸ D3.3.7	• investigate passive solar design, or superinsulation measures (as D3.3.7 and D4.3.3);
▸ D4.3.3	• use recycled (or recyclable) materials and products where possible (as D4.7.8) and timber
▸ D4.7.8	from sustainably managed sources;
	• use products with low embodied energy;
▸ D3.3.3	• review D3.3.3 to reduce reliance on private cars and encourage use of public transport;
	• design heating, lighting and ventilation systems to minimise fossil fuel consumption;
	• select materials of lowest toxicity;
	• detail carefully to ensure airtight construction and elimination of cold bridges;
	• whenever possible avoid halon-based fire control systems;
	• provide for efficient use of daylight in designs for the commercial and institutional sectors.

Good practice references and further reading:	• *Renewable energy in the UK: the way forward*, Energy Paper No.55, HMSO, London, 1988.
	• Curwell, S.R., March, C.G. and Venables, R.K., (Eds), *Buildings and Health, The Rosehaugh Guide to the Design, Construction, Use and Management of Buildings*, RIBA Publications, 1990.
	• Building Research Establishment, *CFCs and buildings*, Digest 358, 1991 (New Edition 1992).
	• BRE Information Paper PD8/93: *Guidance on the Phase-Out of CFCs for owners and operators of air conditioning systems*, January 1993.
	• Department of Trade and Industry, *CFCs and Halons, Alternatives and the scope for recovery for recycling and destruction*, DTI, HMSO, 1990.

Issue D4.7.5	Sourcing of timber

Background:	There is evidence to show that timber can offer energy savings over energy-intensive materials such as brickwork and aluminium. During its processing and use in construction, timber creates less pollution and lower energy inputs than other materials. Excluding transport energy, a solid timber beam has one nineteenth the embodied energy of a steel I-beam equivalent and a 'glulam' beam one sixth of its equivalent. However, more widespread use of timber raises the issue of deforestation. Britain is the seventh largest consumer of tropical hardwood in the developed world, and it is estimated that the construction industry accounts for around 50% of the UK timber consumption.
	The ultimate goal must be use of timber only from sustainably managed sources. However, for tropical hardwoods, such sources are not yet extensive (a recent study for the International Tropical Timber Organisation found that less than 0.2% of tropical moist forests are being managed sustainably for commercial production) so tropical timber should only be used for those purposes for which there are currently no other timbers or wood products with comparable technical properties, or other materials that do not pose a greater environmental hazard.
	Various interest groups in the timber trade are taking steps in the right direction and a number of separate initiatives are under way though Friends of the Earth and World Wide Fund for Nature remain active and forceful in their campaigns about tropical de-forestation. A number of interested parties are currently working to establish the Forest Stewardship Council to accredit organisations to certify forests as well managed and sustainable.
	The Timber Trade Federation (TTF) has initiated an import surcharge on tropical hardwoods to raise funds for forest management and conservation. However, the rate of reforestation is only 1 million ha per year, around 7% of what is being removed.
	One of the problems for specifiers is that it is countries rather than timber species that have sustainable management and reliable information is hard to obtain. The establishment of the Forest Stewardship Council later in 1993 is aimed at providing specifiers and buyers with a means of identifying suitable sources and other initiatives aimed at certifying well-managed forests are being developed.

Background references:	• Construction Industry Environmental Forum, *Purchasing and specifying timber*, Notes of meeting held on 23/02/93, CIRIA, 1993. • Johnson, B., *Responding to Tropical Deforestation – An eruption of crises – An array of solutions*, An Osborn Center Research Paper (WWF), 1991. • Dudley, N., *Forest in Trouble: A Review of the Status of Temperate Forests Worldwide*, WWF, 1992. • The Forest Stewardship Council, A Discussion Paper, 1993.

Good Practice:	*To ensure as far as possible that the timber specified comes from a well-managed source:* • follow developments in the establishment of the Forest Stewardship Council and other certification schemes; • seek information from timber suppliers on the source of the timber they offer and the environmental policies of the forest owner or manager; • use timber suppliers that have agreed to stop selling or specifying non-sustainably produced tropical timber or timber products, or have alternatives available; • for technical and structural properties of alternative timbers, consult BRE Timber Division or TRADA; • consult *The Good Wood Manual* and *Timber: Types and sources* which provide guidance on the timbers to avoid and acceptable sources of supply; • consider substitutes for tropical hardwoods and tropical plywoods, such as temperate hardwoods, softwoods, or other materials – the use of lesser-known species both takes the pressure off over-exploited species, and encourages more intensive felling in areas where there is a wide range of species growing; • consult the *Greener Building Products and Services Directory*, Section 4; • consider the use of reclaimed timber wherever possible.

Good practice references and further reading:	• Hall, K and Warm, P., *Greener Building Products and Services Directory*, AECB, 1993. • *Timber: Eco-labelling and Certification*, Publication L295, Friends of the Earth, 1993. • *Timber: Types and sources*, Publication L296, Friends of the Earth, 1993. • *The Good Wood Manual*: Specifying Alternatives to Non-renewable Tropical Hardwoods, Friends of the Earth, January 1990. • *Architects Journal* : Tropical Hardwoods, Fruits of the Forest, 8 August 1990. • *Sustainability and the Trade in Tropical Rainforest Timber*, Friends of the Earth, undated. • Building Research Establishment, *A Handbook of Softwoods*, SO 39, 1977. • Building Research Establishment, *Timber for joinery*, Digest 321, 1987. • Building Research Establishment, *Handbook of Hardwoods*, SO 7, 1972. • Building Research Establishment, *Timbers: their natural durability and resistance to preservative treatment*, Digest 296, 1985. • Construction Industry Environmental Forum, *The use of timber in construction*, Notes of a meeting held on 23/9/93, CIRIA, 1993.

Issue D4.7.6	**Transport of materials and components**
Background: ▸ D3.3.3 ▸ D3.3.4 ▸ D4.7.3 ▸ D1 ▸ D2	Approximately 75% of all commercial energy consumed by human activity is derived from finite sources of fossil fuels. Burning this fuel releases pollutants into the atmosphere which adds to global warming, as well as leading to other problems such as acid rain, photo-chemical smogs etc. Although the majority of the energy produced from fossil fuels is consumed by the occupants of finished buildings and structures, some of that energy is used in the transportation of construction materials and equipment to the site. About half the energy consumed in this country is used in buildings, the remainder being equally split between transport and industry (*Greener Buildings Products and Services Directory*). For further background on this issue, refer to D4.7.3: Criteria for selection. It is not possible to give firm data regarding transport and the construction industry, other than the general principle that transport should be reduced where possible. This applies to: • choice of fuel; • choice of materials; • siting of each project (see Stages D1 and D2 in this Handbook). It is difficult to foresee any significant impact on choice of fuel or siting of the building without recourse to regulation or legislation in the industrial free market economies. Although transport makes a contribution to energy use, it remains far less significant than the contribution from conservation of direct energy consumed in buildings.
Background references:	• Hall, K and Warm, P., *Greener Building Products and Services Directory*, Association for Environment Conscious Building Directory, Second Edition, 1993. • Pearson, D., *The Natural House Book*, Gaia Books, Conran Octopus, London, 1989.
Good Practice: ▸ D4.7.3	*When considering the transport implications on a project:* • select materials from a local source wherever possible (especially if it is necessary to import an alternative from another country); • review D4.7.3 for good practice guidance on criteria for selection of materials (embodied energy); • investigate options for transporting materials, components and sub-assemblies to site by water.
Good practice references and further reading:	• See background references, plus: • As D4.7.3. • See D3.3.3 and D3.3.4.

Issue D4.7.7	Vetting suppliers
Background:	There is an information gap when it comes to weighing the relative merits and disadvantages of the properties of different building materials and especially of suppliers. We have systems to classify most materials in terms of strength, density, weight, durability etc. (physical characteristics) and all materials have to be tested and comply with the standards of an official body, such as the British Standards Institution or British Board of Agrément. In addition, much emphasis has been placed on current legislation on requiring manufacturers to supply safety information with products. But there is a lack of information generally on the environmental impact of materials, particularly for synthetic products. Suppliers generally fail to give such information, and may not know it themselves. Specifiers will find it almost impossible to ascertain how much energy has been used in the production and processing of materials and other data, such as sources of hardwoods, the amount that has been recycled in paper and metal products etc. Manufacturers and suppliers have a clear responsibility for providing this information in the same way that information is now available in the food industry.
	'Consumers make known their opinions of products by exercising choice. They are increasingly interested not only in the product itself but also in the various raw materials and processes by which it is made' (Pearson, *The Natural House Book*, p.129).
▸ D3.8.2	Refer to D3.8.2 for further background on the green labelling of materials and products. In particular, this refers to the likely amendments to the EC Dangerous Substances Directive, requiring labelling substances dangerous to the environment. Also, the EC Regulation (880/92) setting up an Eco-labelling scheme for product groups has been agreed and launched, the purpose being to impose uniformity on 'environmentally friendly' claims made by manufacturers.
▸ D4.7.5	There are some moves towards vetting of suppliers. • The EC Eco-labelling scheme (UK Ecolabelling Board) as referred to above. • The timber industry has made efforts to vet suppliers. Refer to D4.7.5 for details of the Timber Trade Federation. Friends of the Earth initiated the Good Wood Seal of Approval which they award to companies that have agreed to stop selling or specifying non-sustainably produced tropical timber or timber products. • In 1977 the West German government introduced a scheme under which safe materials could display the 'Blue Angel' symbol. Private laboratories carry out independent tests before granting this symbol. A similar system is needed in the UK, similar to the BSI Kite Mark. • The Association for Environment Conscious Building have produced the *Greener Building Products and Services Directory*. Its aim is to help specifiers to select products that have the least impact on the environment. It is in loose leaf format for ease of updating and has an Appendix containing addresses of all the firms listed in the Directory. • A few building contractors, including some major firms, are now developing environmental policies in terms of selection of materials and site procedures.
	The construction industry must support developments in labelling schemes for building products and appliances. Such labelling is likely to become increasingly important and may become mandatory through EC Directives.

Issue D4.7.7	**Vetting suppliers**

Background references:	• Hall, K and Warm, P., *Greener Building Products and Services Directory*, Association for Environment Conscious Building Directory, Second Edition, 1993. • Pearson, D., *The Natural House Book*, Gaia Books, Conran Octopus, London, 1989. • Evans, B., Rating environmental impact, Article in *Architects' Journal*, 19 May 1993. • Ove Arup & Partners, *The Green Construction Handbook – A Manual for Clients and Construction Professionals*, JT Design Build, Bristol, 1993.
Good Practice: ▸ D4.7.5 ▸ D3.8.2	***When vetting suppliers, consider the following actions:*** • for timber products and suppliers, review the good practice guidance in D4.7.5 and implement relevant measures; • review good practice guidance in D3.8.2 and seek the relevant information from suppliers; • investigate options in materials, equipment and suppliers, demand information on their environmental policy and select only those that have the least impact on the environment; • refer to *Greener Buildings, Products and Services Directory* for information on suppliers and use firms listed in the Appendix if possible; • refer to *Green Design* which lists building materials in alphabetical order with an Appendix of useful addresses for organisations to contact, suppliers, etc; • contact the UK Ecolabelling Board for specific advice on products and manufacturers.
Good practice references and further reading:	• Construction Industry Environmental Forum, *Environmental issues in construction – A review of issues and initiatives relevant to the building, construction and related industries*, CIRIA Special Publications 93 and 94, 1993, Sections 3 and 4. • *Environmental Labelling*, London DTI (DOE/DTI Quarterly Publication). • EC Ecolabelling Scheme: Guidelines for Business, CBI, 1993. • Hall, K and Warm, P., *Greener Building Products and Services Directory*, Association for Environment Conscious Building Directory, Second Edition, 1993. • *Buying into the environment – Guidelines for integrating the environment into purchasing and supply*, Business in the Community, London, 1993. • Elkington, J., Knight, P., Hailes, J., *The Green Business Guide – How to take up – and profit from – the environmental challenge*, Victor Gollancz, London 1991. • Fox, A. and Murrell, R., *Green Design: A guide to the environmental impact of building materials*, Architecture Design and Technology Press, 1989. • Contact: UK Ecolabelling Board: 071-820 1199.

Issue D4.7.8	**Use of recycled materials**
Background: ▸ D3.13.1	Until the development of extended transport systems, and relatively cheap fuel, building materials were usually of local origin and readily available. In addition they required little processing before use in construction. In most cases, this also meant that they were easily re-used. The materials we use affect the environment – the winning of raw materials, and their transportation both deplete resources and consume energy. More and more building materials are becoming scarce, and if present trends continue, some of the most common raw materials will be exhausted during the next 30–40 years. 'Traditional materials – clay, lime, chalk, stone – still abound, and timber (especially softwoods) can be replenished by properly managed forestation. In addition, these materials are easily reused or recycled, they produce little or no pollution, and they are reabsorbed into the natural cycles of the environment once their use as building materials is over'. (David Pearson, *The Natural House Book*, p.128.) Review D3.13.1 and relevant background references for useful information at design Stage 3. A major constraint to greater recycling (particularly bulk materials) is the transport costs compared to the value of the materials. Davis Langdon and Everest (see CIRIA SPs 93 & 94) have estimated that the demolition industry recycles about 11 million tonnes of the construction industry's waste (from structures that have reached the end of their life). But there is potential for recycling of a further 70% of the material currently going to landfill.
Background references:	• Pearson, D., *The Natural House Book*, Gaia Books, Conran Octopus, London, 1989. • Ove Arup & Partners, *The Green Construction Handbook – A Manual for Clients and Construction Professionals*, JT Design Build, Bristol, 1993. • Construction Industry Environmental Forum, *Environmental issues in construction – A review of issues and initiatives relevant to the building, construction and related industries*, CIRIA Special Publications 93 and 94, 1993, Section 4.
Good Practice: ▸ D3.13.1	*When considering the potential use of recycled material take as many of the following steps as are appropriate to the project.* • CFCs from existing air conditioning and refrigeration equipment must be recycled and not released to the atmosphere. • Review D3.13.1 and implement any appropriate measures. • Select products that are capable of being reused and recycled. This can save large amounts of energy spent on processing raw materials. Recycled steel, for example, saves more than 70% of the energy used in manufacturing new steel from primary ore. • Use salvaged materials, such as doors, beams, stone, bricks, tiles and slates which is an environmental option even cheaper than recycling materials. In 1977, the *Architects' Journal* promoted a new service, the Architectural Salvage Scheme, to publicise the availability of suitable items. This service is still operating, but for more specialist items. • Select good quality materials (to ensure longevity). Terms such as 'built-in obsolescence' or 'disposable' are anathema to any environmental policy. • Dry construction techniques are to be preferred to wet construction. Concrete for example, cannot be recycled (other than as hardcore, or possibly as a secondary aggregate). Screed floors in flats and houses cannot be recycled, whereas timber floors on battens as an equivalent construction allows for re-use. Modern cement-rich mortars effectively preclude re-use of bricks which cannot be salvaged intact, whereas lime-rich mortars allow for easy re-use of bricks. Nevertheless, there is a well-developed market for second-hand bricks, amounting to 75 to 100 million by number in 1986, representing 2% of annual brick consumption.

Issue D4.7.8	Use of recycled materials

Good Practice continued:	• Where options exist, design for flexibility and adaptability. Many Georgian and Victorian speculative terraced houses have been converted successfully into offices, or flats, and many people believe that these are more desirable than the purpose built modern equivalent.
	• Aggregates (sand and gravel) have been thought of as freely available, but extraction is becoming increasingly difficult from various sensitive sites. Extraction is environmentally destructive (even if sea-dredged), and alternatives, such as crushed rock, or secondary aggregates (from demolished buildings) must be considered.
	• The increased use of performance specifications, partly due to design and build procurement methods, does allow greater flexibility in the choice of materials.
	• Specifications to ensure quality, consistency and correct application could contribute to promoting further recycling.
	• Use products that have been produced from waste materials. At least one company is producing an insulation material from old newsprint treated for fire-resistance and mould prevention, though it is presently only appropriate for roof insulation. Shredded rubber from car tyres can be used in some construction products such as sports surfaces, carpet underlay, road asphalt and print additives. Research at the Transport Research Laboratory has recently led to the introduction of recycled materials into the UK Specification for Highway Works.
	• Provide storage space for recycling of waste from the building's occupants. Offices dispose of enormous quantities of waste paper and card packaging. Bottles can be easily recycled (about 30 to 35% of glass is recycled from refuse). Other items that may be recycled from waste from buildings in use are:
	  – Catering waste – organic and non-organic;
	  – Plastics;
	  – Industrial waste – metallic and plastic;
	  – Chemical waste and effluents;
	  – Timber.

Good practice references and further reading:	• See background references, plus: • Fox, A. and Murrell, R., *Green Design: A Guide to the Environmental Impact of Building Materials*, Architecture Design and Technology Press, 1989. • Construction Industry Environmental Forum, *Environmental issues in construction – A review of issues and initiatives relevant to the building, construction and related industries*, CIRIA Special Publications 93 and 94, 1993. • Construction Industry Environmental Forum, *Recycling on site – the practicalities*, Notes of a meeting held on 22 June 1993, CIRIA, 1993. • Construction Industry Environmental Forum, *Recycling in construction: The use of recycled materials*, Notes of a meeting held on 13/7/93, CIRIA, 1993. • Edwards, A.C. and Mayhew, H.C., *Recycled asphalt wearing courses*, Transport Research Laboratory Report RR225, 1989. • Cornelius, P.D.M. and Edwards, A.C. *Assessment of the performance of off-site recycled bituminous material*, Transport Research Laboratory Report RR305, 1991. • *Monitoring of cold road recycling process on a heavily-trafficked road*, ETSU New Practice Report NP/60, 1992.

D4.8   End-use considerations

Issue D4.8.1	Environmental issues in respect of end-use
Background:	During the detail design stage it is important to review primary design decisions in relation to policy and briefing requirements and end-use considerations. In purpose-designed buildings, there will be opportunities to review, with the client and representatives of building users, end-use considerations at each stage in the design process.
	In speculative buildings the brief is likely to be aimed at producing a marketable product, which in commercial sector buildings may entail minimum circulation and shared areas, and possibly air conditioning if the building is for a prestige market. Wherever possible the designer should stress the importance of environmental effects and end-use considerations, particularly in relation to subsequent fitting out.
	Opportunities for waste management and provision of storage space for recyclable materials also need to be reviewed and finalised at this stage.
Background references:	No specific background references identified.
Good Practice: ▸ D3.13.1 ▸ D4.7.8	*In designing for end-use requirements:* • wherever possible involve users and occupants in design at the earliest possible stage and review in particular requirements for heating, lighting and ventilation with them before completing detail design; • discuss resource, energy and water use targets with clients and their building managers or tenants; • consider how space will be let and fitted out in speculative commercial or industrial buildings and take this into account in space planning and services provisions; • review waste disposal and provisions for storage for recyclable materials (see D3.13.1, D4.7.8).
Good practice references and further reading:	• Halliday, S.P., *Environmental Code of Practice for Buildings and Their Services*, BSRIA, 1994. • Vale, B., Vale, R., *Towards a Green Architecture*, Six Practical Case Studies, RIBA Publications, 1991, Case Study 5, Pages 58–59.

Issue D4.8.2	Operational and post-operational requirements
Background:	How a scheme will be managed and how occupants and/or users will operate the systems and controls and maintain it are increasingly recognised as vital elements in its energy efficient running and environmental friendliness. In addition, the amount of control users and occupants can exercise in gaining access to fresh air, warmth and natural light may have a significant impact on their wellbeing and productivity. Complicated control systems can in themselves be a problem unless they are designed with provision for good monitoring and management and are sympathetic to user needs.
Background references:	• Halliday, S.P., *Environmental Code of Practice for Buildings and Their Services*, BSRIA, 1994. • See D3.13.4 in this publication for background information on design for maintenance and cleaning.
Good Practice: ▸ D3.13.4	*In designing for operational and post-operational requirements:* • review the handbook on good practice in design for maintenance and cleaning (see D3.13.4) and ensure that adequate provisions are made in working up the detail design and specification; • stress to clients the importance of competent service personnel where necessary and assigning adequate resources for operation and maintenance; • provide generous access space for ease of leak checking and recovery of refrigerants if air conditioning is necessary; • match the control regime to the intended usage or tenancy pattern, and consider the appropriateness of building management systems where complicated services installations are required; • select least polluting energy sources and consider possible changes in use; • include a requirement for contractors and services contractors to provide details of all materials employed, details of control systems and their operations, and recommended maintenance periods and procedures, for inclusion in a user or tenant manual.
Good practice references and further reading:	• See background references, plus: • *Operating and Maintenance Manuals for Building Services Installations*, BSRIA AG1, 1987. • *Maintenance Management for Building Services*, CIBSE, TM 17, 1990. • *Building Services Maintenance*, BSRIA, RG No.1, Vol 1 and 2, 1990. • *Standard Maintenance Specification for Mechanical Services in Buildings*, HVCA. • *Decision in Maintenance*, BSRIA, TN14/92, 1992.

Issue D4.8.3	Commissioning

Background:	The importance of commissioning a scheme, and its management (operation and maintenance) are vital to achieving energy efficiency in use. The design engineer can make a significant impact on this energy efficiency by simplifying the commissioning and management process, considering how systems will be operated, and by selecting and specifying systems accordingly. It is important to allow sufficient time and a suitable budget for specialist personnel (if appropriate) to commission the scheme.
	It is useful to involve the users, including any building occupants, in this commissioning process so that they have a complete understanding of the principles involved and controls provided.
	References to system design and the importance of maintenance and commissioning procedures have been included in the following issues:
▸ D4.3.1 ▸ D4.3.2 ▸ D4.6.1 ▸ D4.6.8 ▸ D4.6.9	• D4.3.1 Heating • D4.3.2 Cooling • D4.6.1 Services • D4.6.8 Legionnaires' Disease • D4.6.9 Sick Building Syndrome.
	All these should be reviewed and any relevant measures incorporated.
Background reference:	• Halliday, S.P., *Building Services and Environmental issues – The Background*, BSRIA Interim Report, April 1992.

Good Practice:	• Design commissionable systems:
	  – consider the appropriateness of a scheme management and information system and keep it simple;
	  – include suitable test points;
	  – provide suitable access for maintenance, installation, commissioning and replacement.
	• If more than one discipline is involved, for example in sewage treatment works, power stations, railways or major buildings, ensure all are involved in a team approach to the commissioning task.
	• Provide full 'user friendly' commissioning codes and any other information on the design to facilitate commissioning.
	• Ensure that the specification calls for a programme on site where commissioning is complete before occupation.
	• Ensure that the specification sets out full performance testing procedures and standards, in a clear unambiguous way.
	• Review other relevant issues (see background references) and carry out revisions if appropriate).
	• If air conditioning is required consider the following:
	  – use equipment designed for low losses or leakage of gases;
	  – specify recyclability of charge;
	  – specify leak detection alarms for CFCs and make provision for abstracting refrigerant from systems.
	• In fire systems, use integrity testing instead of gas dumping to save halons and CO_2 dumping.

Good practice references and further reading:	• Halliday, S.P., *Environmental Code of Practice for Buildings and Their Services*, BSRIA, 1994. • *Pre-commissioning cleaning of water systems*, BSRIA, AG8/91. • *The Commissioning of Water Systems in Buildings*, BSRIA, AG2/91. • *The Commissioning of VAV Systems in Buildings*, BSRIA, AG1/91. • *The Commissioning of Air Systems in Buildings*, BSRIA, AG3/91. • CIBSE, *Commissioning Codes: A: Air distribution Systems (1971); B: Boiler Plant (1975); C: Automatic Controls (1973); R: Refrigerating systems (1991); W: Water distribution systems (1989)*, Chartered Institution of Building Services Engineers, London.

D4.9 Handover of project environmental policy to contractor and/or other consultants

Issue D4.9	**Handover of project environmental policy to contractor and/or other consultants**
Background: ▸ D1.3.2 ▸ D1.4 ▸ C13.1 ▸ C13.2	The Project Environmental Policy is the foundation for tackling all environmental aspects of the project. Any contractors or consultants need to be aware not only of its contents but also the implications for their input and work towards the project's successful realisation. Misunderstandings and poor communication could compromise the earlier commitment made to environmental issues. The project's environmental representatives in the design teams will be directly responsible for ensuring that consultants and/or contractors are fully aware of the implications of the project's environmental policy.
Background references:	See good practice references.
Good Practice:	• The Project Environmental Policy should be made available to all involved in the project in a form which is attractive to read and readily understood. • Where appropriate, full reasoning and explanation behind the policy statements should be included in any documentation provided. • Briefing and training sessions should be organised to ensure full understanding of the implications of the environmental policy and what has been achieved thus far, and relevant papers, reports etc. should be exchanged. • A commitment from the consultants and/or contractors to following the project's policy should be obtained. • The project environmental representative should ensure that a direct counterpart is employed by other consultants or the contractor(s) with whom he or she can communicate and rely upon to ensure policy commitments are followed through.
Good practice references and further reading:	• *Design Briefing Manual*, AG1/90, BSRIA, 1990. • Barwise, J., and Battersby, S., *Environmental Training*, Croner Publications, 1993. • BS7750: 1992 *Specification for Environmental Management Systems*, British Standards Institution, Milton Keynes. • CBI, *Corporate environmental policy statements*, Confederation of British Industry, London, June 1992. • Halliday, S.P., *Environmental Code of Practice for Buildings and Their Services*, BSRIA, 1994. • Construction Industry Environmental Forum, *Environmental Management in the Construction Industry*, Notes of meeting held on 22/09/92, CIRIA, 1992. • Miller, S., *Going Green*, JT Design Build, Bristol, undated.

Stage D5 Environmental considerations at tendering and contract letting

Stage D5: Environmental considerations at tendering and contract letting covers the processes of drawing up documents for issue with invitations to tender, the development of appropriate select tender lists of potential contractors, and the selection by clients and their professional advisors of the most appropriate contractor. Contractors' concerns and responsibilities at tender stage are dealt with in Stage 1 of the Construction Handbook. (Venables, R.K et al, *Environmental Handbook for Building and Civil Engineering Projects, Volume 2: Construction Phase*, CIRIA Special Publication 98, 1994).

Issue D5	Environmental considerations at tendering and contract letting
Background:	As the environmental considerations in new building and civil engineering projects, both explicit and implied, become more and more important, increasing use of special terms covering environmental issues in contract documents can be expected. At present, most contract documents only include explicit environmental issues as part of the specification, for example in making specific provisions for the exclusion of CFCs or particular solvents, in specifying ways in which particular legal requirements are to be met, or in specifying particular 'environmentally-friendly' materials. However, general environmental provisions, such as a total ban on the use of CFCs in a project, or the use only of diesel-engined vehicles, continue to be slowly introduced, although few in number at present. Such provisions are anticipated to become commonplace as environmental management systems based on BS7750 are developed and clients come to expect contractors to have such systems in place. This development is forecast by some to mirror the development of quality management systems based on BS5750, with some clients requiring any contractor wishing to tender to have established a certified quality management system or a total quality management system to BS7850. However, since BS7750 is currently being revised, only the most environmentally conscious contractors will have simple BS7750-derived environmental management systems in place now (late-1993). Accredited certification is not likely to be available until mid-1994. Separately, as the earlier sections have demonstrated, the legal and regulatory environmental requirements are increasing. There is a need for clients and their professional advisors to explain to prospective contractors the main environmental facets of the project and the legal requirements contractors will be expected to meet. In addition, if a formal environmental policy for the project has been agreed, or formal statements of specific environmental requirements drawn up, it will be appropriate or even necessary to include them in tender documents and formally require contractors when tendering to make specific provision for meeting the policy requirements.
Background references:	• Miller, S., *Going Green*, JT Design Build, Bristol. • Construction Industry Environmental Forum, *Environmental issues in construction – A review of issues and initiatives relevant to the building, construction and related industries*, CIRIA Special Publications 93 and 94, 1993. • CIC Environment Task Group, *Our land for our children: an environmental policy for the construction professions*, Construction Industry Council, August 1992. • CIEC Environment Task Force, *Construction and the Environment*, Building Employers Confederation, May 1992. • BS7750: 1992 *Specification for Environmental Management Systems*, British Standards Institution, Milton Keynes. • BS5750: 1987 *Quality Systems*, British Standards Institution, Milton Keynes. • BS7850: 1992 *Total Quality Management*, British Standards Institution, Milton Keynes.
Good Practice:	*In drawing up tender documents:* • include a section called Environmental Issues and Policies containing any agreed environmental policy statements, any specific environmental requirements and explicit explanations of what is required of the tenderer to meet them; • whilst requiring all contractors to comply with all current legislation that is applicable to the work and without removing such a duty from the contractors, consider including a list of relevant environmental legislation which the client wishes to draw especially to the tenderers' attention; • consider drawing attention in the general provisions of the tender documents to common environmental features of items in the specification, such as exclusion of CFCs or the use of temperate rather than tropical hardwoods, which would otherwise only be discovered by tenderers as they studied the details; • draw attention to particular environmental features, sensitive areas or potential impacts of the project; • include all relevant site investigation data in the tender documents; • if appropriate, include guidance on how contractors should demonstrate the commitment sought, for example participation in considerate contractor schemes and ask for a statement of proposed modifications to, or inability to comply with, the environmental requirements.

Issue D5	Environmental considerations at tendering and contract letting

Good Practice continued: ▸ D2.3.9 ▸ D1.4	***In drawing up a select tender list:*** • consider, as environmental management systems are developed, the extent to which the client wishes to restrict his or her choice of tenderer to those firms that have an environmental management system in place or can in some other way demonstrate their commitment to environmentally responsible operations; • consult the client(s) on their wishes on this issue; • consider adding environmental performance to the normal selection criteria of quality, financial position, and completion on time and to budget • consider seeking from potential tenderers a copy of their corporate environmental policy statement, method statements and/or other documentation that demonstrates their commitment to environmentally responsible operations; • consider securing information on considerate contractor schemes, the eco-labelling scheme and BREEAM and adding participation in such schemes to your list of environmental criteria on which contractors may be selected. ***In comparing tenders on environmental grounds and making a final selection:*** • confirm the client's view of the influence environmental considerations should have on the final selection so that the selection criteria are clear before the comparison is started; • in particular, agree whether the client is prepared to accept a price premium for engaging what is judged to be a more environmentally responsible contractor; • review the documentation on environmental performance submitted by the tenderers and rank them in environmental priority order for comparison with the other selection criteria rankings; • in that review, consider the relative importance to you and your client of, on the one hand, past actual performance of a stated environmental policy which may be limited in scope, and on the other, policy statements which may display a stronger environmental commitment but which have not yet been put into practice. ***When agreeing the contract with the contractor:*** • ensure that any environmental requirements included in the tender documents are transferred to the contract documents; • similarly, ensure that any environmental issues included in their tender by the successful contractor are also transferred to the contract documents; • check whether any additional environmental requirements have come to light since tender invitations were issued and consider whether and how they are to be covered in the contract; • if not already agreed, ensure agreement on measures for environmental monitoring of the contractor's work during the contract; • ensure the appointment by the contractor of an individual responsible for environmental liaison and/or the role of Environmental Manager as envisaged by BS7750.
Good practice references and further reading:	• BS7750: 1992 *Specification for Environmental Management Systems*, British Standards Institution, Milton Keynes. • Atkinson, C.J., and Butlin, R.N, *Ecolabelling of building materials and building products*, BRE Information Paper 11/93, BRE, 1993. • Prior, J.J., (Editor), BREEAM 1/93, *An environmental assessment for new offices*, BRE, 1993. • Prior, J.J., Raw, G.J. and Charlesworth, J.L., BREEAM 3/91, *An environmental assessment for new homes*, BRE, 1991. • Baldwin, R., Bartlett, P., Leach, S.J. and Attenborough, M., BREEAM 4/93, *An environmental assessment for existing office buildings*, BRE, 1993. • No references have been identified as giving specific guidance on how to include environmental matters in tender documents for building and civil engineering projects. However, the following guide gives cogent guidance on inclusion of health and safety requirements which will in time be extendable to environmental matters. • European Construction Institute, *Total project management of construction safety, health and environment*, Thomas Telford, 1992.
Legal references:	No specific legal references have been identified as giving specific guidance on what environmental matters to include in tender documents for building or civil engineering projects.

List of main background and good practice references

A Books, reports and guidance documents.

Atkinson, C.J., and Butlin, R.N., *Ecolabelling of building materials and building products*, BRE Information Paper 11/93, BRE, 1993.

Baldwin, R., Bartlett, P., Leach, S.J. & Attenborough, M., BREEAM 4/93, *An environmental assessment for existing office buildings*, BRE, 1993.

Barwise, J., and Battersby, S., *Environmental Training*, Croner Publications, 1993.

Bright, K., *Building a greener future – Environmental issues facing the construction industry*, CIOB Occasional Paper 49, Chartered Institute of Building, 1991.

BS7750: 1992 *Specification for Environmental Management Systems*, British Standards Institution, Milton Keynes.

BSI, *BS7750 2nd Edition Draft for Public Comment*, DC400220/93, BSI, London, 1993.

Building Research Establishment, *CFCs in Buildings*, BRE Digest 358, 1992.

Butler, D., & Howard, P.N., Life Cycle CO_2 Emissions: From the Cradle to the Grave, in *Building Services*, 1992.

Butler, D. J. G., *Guidance on the phase-out of CFCs for owners and operators of air conditioning systems*, PD25/93, Building Research Establishment, 1993.

CBI, *Corporate environmental policy statements*, Confederation of British Industry, London, June 1992.

CIBSE, *Building Energy Code*, in 4 parts, CIBSE, 1975–82.

CIC Environment Task Group, *Our land for our children: An environmental policy for the construction professions*, Construction Industry Council, August 1992.

CIEC Environment Task Force, *Construction and the Environment*, BEC, May 1992.

Climate and Site Development, BRE Digest No. 350, in 3 parts, BRE, 1990.

Construction Industry Environmental Forum, Notes of meetings:.

> *Materials and product blacklists*, 23/7/92, CIRIA, 1992.
> *Building lifetimes*, 27/7/92, CIRIA, 1992.
> *Environmental management in the construction industry*, 22/09/92, CIRIA, 1992.
> *Environmental considerations in public sector refurbishment*, 12/10/92, CIRIA, 1992.
> *Water pollution from construction sites*, 20/10/92, CIRIA, 1992.
> *Buildings and energy*, 25/11/92, CIRIA, 1992.
> *Green clients – The role of the client in setting the environmental tone of construction projects*, 30/11/92, CIRIA, 1992.
> *Green buildings: the designer's perspective*, 8/12/92, CIRIA, 1993.
> *Contaminated land*, 12/1/93, CIRIA, 1993.
> *Eco-labelling: the implications for the construction industry*, 26/1/93, CIRIA, 1993.
> *Purchasing and specifying timber*, 23/02/93, CIRIA, 1993.
> *Considerate builders and contractors*, 9/3/93, CIRIA, 1993.
> *The environment, economics and the construction industry*, 19/03/92, CIRIA, 1992.
> *Nature conservation issues in building and construction*, 23/3/93, CIRIA, 1993.
> *Waste management*, 6/4/93, CIRIA, 1993.
> *Life-cycle eco-analysis in building and construction*, 25/5/93, CIRIA, 1993.
> *Re-cycling on site – The practicalities*, 22/6/93, CIRIA, 1993.
> *Recycling in construction – The use of recycled materials*, 13/7/93, CIRIA, 1993.
> *Contaminated land*, 14/9/93, CIRIA, 1993.
> *The use of timber in construction*, 23/9/93, CIRIA, 1993.

Construction Industry Environmental Forum, *Environmental issues in construction – A review of issues and initiatives relevant to the building, construction and related industries*, CIRIA Special Publications 93 and 94, 1993.

Coppin, N.J. & Richards, I.G., *Use of vegetation in civil engineering*, CIRIA Book 10, 1990.

Cost effective management of reclaimed derelict sites, DoE, HMSO, 1989.

Crisp, V.H.C, Doggart, J., and Attenborough, M., BREEAM 2/91, *An environmental assessment for new superstores and supermarkets*, BRE, 1991.

Croners, *Environmental Management*, with quarterly amendment service, Croner Publications Ltd, October 1991.

Curwell, S.R., March, C.G. and Venables, R.K., (Eds), *Buildings and Health, The Rosehaugh Guide to the Design, Construction, Use and Management of Buildings*, RIBA Publications, 1990.

DoE, *The use of halons in the United Kingdom and the scope for substitution*, DOE,, HMSO, 1991.

Department of Trade and Industry, *CFCs and Halons: Alternatives and the scope for recovery for recycling and destruction*, DTI, HMSO, 1990.

Department of Transport, *Design Manual for Roads and Bridges, Volume 10: Environmental Design*, HMSO, 1992.

Department of Transport, *Manual of Environmental Assessment*, HMSO, July 1993.

Design Briefing Manual, AG1/90, BSRIA, 1990.

DoE & DoT, *Reducing transport emissions through planning*, HMSO, April 1993.

Elkington, J., Knight, P., Hailes, J., *The Green Business Guide – How to take up – and profit from – the environmental challenge*, Victor Gollancz, London 1991.

Energy Efficiency Office, *Energy consumption in offices – a technical guide for owners and single tenants*, Energy Consumption Guide 19, EEO, 1991.

Environmental Assessment: A guide to the identification, evaluation and mitigation of environmental issues in construction schemes, CIRIA Special Publication 96, 1993.

Fox, A. and Murrell, R., *Green Design: A guide to the environmental impact of building materials*, Architecture Design and Technology Press, 1989.

Government White Paper: *This Common Inheritance – Britain's Environmental Strategy (1990)* and *Yearly Reports (1991 and 1992)*.

Guidance on the sale and transfer of contaminated land, Draft for open consultation, CIRIA, October 1993.

Hall, K and Warm, P., *Greener Building Products & Services Directory*, Association for Environment Conscious Building Directory, Second Edition, 1993.

Halliday, S.P., *Building Services and Environmental Issues – The Background*, BSRIA Interim Report, April 1992.

Halliday, S.P., *Environmental Code of Practice for Buildings and Their Services*, BSRIA, 1994.

Hooker, P.J. and Bannon, M.P., *Methane: its occurrence and hazards in construction*, Report 130, CIRIA, London, 1993.

Householder's Guide to Radon, 3rd Edition, DoE, 1992.

ICRCL 17/78, *Notes on the development and after-use of landfill sites*, ICRCL, 1990.

Johnston, J. and Newton J., *Building Green – A guide to using plants on roofs, walls and pavements*, London Ecology Unit, 1993.

Leach, B.A., and Goodger, H.K., *Building on derelict land*, CIRIA Special Publication 78, 1991.

Building for the Environment: An Environmental Good Practice Checklist for the Construction and Development Industries, Leicester County Council jointly with Leicester City Council, November 1992.

Miller, S., *Going Green*, JT Design Build, Bristol.

Ministry of Agriculture, Fisheries and Food, *Environmental Procedures for Inland Flood Defence Works – A guide for managers and decision makers in the NRA, Internal Drainage Boards and Local Authorities*, 1992.

Ove Arup & Partners, *The Green Construction Handbook – A Manual for Clients and Construction Professionals*, JT Design Build, Bristol, 1993.

Pearson, D., *The Natural House Book*, Gaia Books, Conran Octopus, London, 1989.

Potter, I.N., *Sick Building Syndrome*, Report BTN04/88, BSRIA, 1988.

Prior, J.J., (Editor), BREEAM 1/93, *An environmental assessment for new offices*, BRE, 1993.

Prior, J.J., Raw, G.J. and Charlesworth, J.L., BREEAM 3/91, *An environmental assessment for new homes*, BRE, 1991.

Shorrock, L.D., Henderson, G., *Energy use in buildings and carbon dioxide emissions*, BR 170, BRE 1990.

Skoyles, E.R. & Skoyles, J.R., *Waste Prevention on Site*, Mitchell Publishing, 1987.

The Environment Council, *Business and Environment Programme Handbook*, 1992 plus regular updates.

Therivel, R. et al, *Strategic environmental assessment*, Earthscan Publications, London, 1992.

Vale, B. and Vale, R., *Green Architecture – Design for a sustainable future*, Thames and Hudson, London 1991

Venables, R.K et al, *Environmental Handbook for Building and Civil Engineering Projects, Volume 2: Construction Phase*, CIRIA Special Publication 98, 1994.

Main references

B Main legislation.

UK

Alkali Works etc Regulation Act 1906.

Building (Scotland) Act 1970.

Building Act 1984.

Building Regulations, 1991.

Clean Air Acts 1956 and 1958.

Control of Pollution Act 1974.

Control of Substances Hazardous to Health Regulations 1988 (SI 1988 No. 1657) (as amended by SI 1990 No. 2026 and SI 1992 No. 2382).

Construction Products Regulations 1991 (SI 1991 No. 1620).

Electricity and Pipe-line Works (Assessment of Environmental Effects) Regulations 1989 (SI 1989 No. 167).

Environmental Protection Act 1990.

Harbour Works (Assessment of Environmental Effects) Regulations 1988 (SI 1988 No. 1336) and (No.2) Regulations 1989 (SI 1989 No. 524).

Health and Safety at Work etc Act 1974.

Highways (Assessment of Environmental Effects) Regulations 1988 (SI 1988 No. 1241).

Land Drainage Improvement Works (Assessment of Environmental Effects) Regulations 1988 (SI 1988 No. 1217).

National Parks and Access to the Countryside Act 1949.

Noise at Work Regulations 1989 (SI 1989 No. 1790).

Occupier's Liability Acts 1957 and 1984.

Planning (Consequential Provisions) Act 1990.

Planning (Hazardous Substances) Act 1990.

Planning and Compensation Act 1991.

Town and Country (Assessment of Environmental Effects) Regulations 1988 (SI 1988 No. 1199) (as amended by SI 1990 No. 367 and SI 1992 No. 1494).

Town and Country Planning (Scotland) Act 1972.

Town and Country Planning Act 1990.

Water Industry Act 1991.

Water Resources Act 1991.

Wildlife and Countryside Act 1981 (as amended).

Workplace (Health, Safety and Welfare) Regulations 1992 (SI 1992 No 3004).

EC

Directive on Construction Products 89/109/EEC.

Directive on Marketing and Use of Certain Dangerous Substances 76/769/EEC (as Directive on the assessment of the effects of certain public and private projects on the environment 85/337/EEC.

Directive on the assessment of the effects of certain public and private projects on the environment 85/337/EEC.

Directive on the Conservation of Wild Birds (79/409/EEC).

Directive on the Conservation of Wild Fauna and Flora (92/43/EEC) (Habitats Directive) (Intended to be implemented by 05.06.1994).

Directive on the minimum safety and health requirements for the workplace (89/655/EEC).

EC Regulation (880/92) on a Community Eco-Label Award Scheme.

Addresses of relevant organisations

Association of Consulting Engineers
Alliance House
12 Caxton Street
LONDON SW1H 0QL

Phone: 071-222 6557   Fax: 071-222 0750

The Association for Environment Conscious Building
Windlake House
The Pump Field
COALEY
Gloucestershire GL11 5DX

Phone: 0453 890757

Arboricultural Advisory and Information Service
Forestry Commission Research Station
Alice Holt Lodge
Wrecclesham
FARNHAM
Surrey GU10 4LH

Phone: 0794 68717

Arboricultural Association
Ampfield House
Ampfield
ROMSEY
Hampshire SO5 9PA

Phone: 0794 68717

Architectural Cladding Association
60 Charles Street
LEICESTER
Leicestershire LE1 1FB

Phone: 0533 536161   Fax: 0533 514568

Association of British Plywood and Veneer Manufacturers
Riverside Industrial Estate, Morson Road
Ponders End
ENFIELD
Middlesex EN3 4TS

Phone: 081-804 2424

Association of Building Component Manufacturers
61-63 Rochester Road
AYLESFORD
Kent ME20 7BS

Phone: 0622 715577

British Aggregate Construction Materials Industries
156 Buckingham Palace Road
LONDON SW1W 9TR

Phone: 071-730 8194   Fax: 071-730 4355

Environment Task Group
Building Employers Confederation
82 New Cavendish Street
LONDON W1M 8AD

Phone: 071-580 5588   Fax: 071-631 3872

British Association of Landscape Industries
Landscape House
9 Henry Street
KEIGHLEY
West Yorkshire BD21 3DR

Phone: 0535 606139

British Earth Sheltering Association
Caer Llan Barn House
Lyddant
MONMOUTH NP5 4JJ

British Library Environmental Information Service
25 Southampton Buildings
Chancery Lane
LONDON WC2A 1AW

Phone: 071-323 7955   Fax: 071-323 7954

British Timber Merchants Association (BTMA)
Stocking Lane
Hughenden Valley
HIGH WYCOMBE HP14 4JZ

Phone: 0494 563602

British Standards Institution
2 Park Street
LONDON W1A 2BS

Phone: 071-629 9000

Enquiry and Ordering Department
British Standards Institution
Linford Wood
MILTON KEYNES
Bucks MK14 6LE

Phone: 0908-226888   Fax: 0908-322484

British Wood Preserving Association
Building No.6
The Office Village
4 Romford Road
Stratford
LONDON E15 4EA

Phone: 081-519 2588

Building Environment Performance Analysis Club
Building Research Establishment
Garston
WATFORD WD2 7JR

Phone: 0923 664138   Fax: 0923 664780

Building Research Energy Conservation Support Unit (BRECSU)
Building Energy Efficiency Division
Building Research Establishment
Garston
WATFORD WD2 7JR

Phone: 0923 664258   Fax: 0923 664787

Addresses of relevant organisations

Energy and Environment Group
Building Services Research and Information
Association
Old Bracknell Lane West
BRACKNELL
Berkshire RG12 4AH

Phone: 0344 426511   Fax: 0344 487575

Environmental Assessment and Futures Section
Building Research Establishment
Garston
WATFORD
Herts WD2 7JR

Phone: 0923 664174 DL   Fax: 0923 664088

Business in the Environment
8 Stratton Street
LONDON
W1X 5FD

Phone: 071-629 1600   Fax: 071-629 1834

Environment Management Unit
Confederation of British Industry
Centre Point
103 New Oxford Street
LONDON WC1A 1DU

Phone: 071-379 7400   Fax: 071-240 0988

Chartered Institution of Building Services Engineers
Delta House
222 Balham High Road
LONDON SW12 9/BS

Phone: 081-675 5211   Fax: 081-675 5449

Construction Industry Council
26 Store Street
LONDON WC1E 7BY

Phone: 071-637 8692

Chartered Institute of Building
Englemere
Kings Ride
ASCOT
Berks SL5 8BJ

Phone: 0344 23355   Fax: 0344 23467

CIRIA (Construction Industry Research &
Information Association)
6 Storey's Gate
Westminster
LONDON SW1P 3AU

Phone: 071-222 8891   Fax: 071-222 1708

Construction Industry Environmental Forum
CIRIA
6 Storey's Gate
Westminster
LONDON SW1P 3AU

Phone: 071-222 8891   Fax: 071-222 1708

Commission of the European Communities
8 Storey's Gate
LONDON SW1P 3AT

Phone: 071-973 1992   Fax: 071-222 0900

Energy Efficiency Office
Department of the Environment
1 Palace Street
LONDON SW1E 5HE

Phone: 071-238 3094   Fax: 071-238 3733

Construction Sponsorship Directorate
Department of the Environment
Room P1/113A
2 Marsham Street
LONDON SW1P 3EB

Phone: 071-276 6728   Fax: 071-276 3826

Dry Lining and Partition Association
82 New Cavendish Street
LONDON W1M 8AH

Phone: 071-580 5588

Energy Design Advice Scheme
The Bartlett Graduate School of Architecture
Philips House
University College, London
Gower Street
LONDON WC1E 6BT

Phone: 071-916 3891

Energy Division
Department of Trade and Industry
1 Palace Street
LONDON SW1E 5HE

Phone: 071-238 3000/3370 Fax: 071-834 3771

Energy System Trade Association
P O Box 16
STROUD
Gloucestershire
GL6 9YB

Phone: 0453 886776   Fax: 0453 885226

Environmental Enquiry Point
Department of Trade and Industry
Warren Spring Laboratory
Gunnels Wood Road
STEVENAGE
Herts SG1 2BX

Phone: 0800 585794   Fax: 0438 360858

The Business and Environment Programme
The Environment Council
21 Elizabeth Street
LONDON SW1W 9RP

Phone: 071-824 8411   Fax: 071-730 9941

European Federation of Waste Management
Avenue des Nerviens 117/69
B-1040 BRUSSELS
BELGIUM

Phone: 010 32 2 732 1601   Fax: 010 32 2 734 9279

Federation of Civil Engineers Contractors
Cowdray House, 6 Portugal Street
LONDON WC2A 2HH

Phone: 071-404 4020   Fax: 071-242 0256

Fencing Contractors Association
St John's House
WATFORD
Hertfordshire WD1 1PY

Phone: 0923 248895

Forest Stewardship Council
c/o WWF UK
Panda House
Weyside Park
Catteshall Lane
GODALMING
Surrey GU7 1XR

Phone: 0483 426444   Fax: 0483 426409

Friends of the Earth
26/28 Underwood Street
LONDON N1 7JQ

Phone: 071-490 1555

Greenpeace
30-31 Islington Green
LONDON N1 8XE

Phone: 071-354 5100

Her Majesty's Inspectorate of Pollution
Romney House,
43 Marsham Street
LONDON SW1P 3PY

Phone: 071-276 8083   Fax: 071-276 8605

Her Majesty's Stationery Office
London Bookshop
49 High Holborn
LONDON WC1V 6HB

Phone: 071-873 0011

Institution of Civil Engineers
Great George Street
LONDON SW1P 3AA

Phone: 071-222 7722

Institute of Environmental Assessment
The Old School
Fen Road
East Kirby
LINCOLNSHIRE
PE23 4DB

Phone: 0790 763613   Fax: 0790 3630

Institute for European Environmental Policy
3 Endsleigh Street
LONDON WC1H 0DD

Phone: 071-388 2117   Fax: 071-388 2826

International Union for Conservation of Nature and Natural Resources
World Conservation Monitoring
Centre
219c Huntington Road
CAMBRIDGE CB3 0DL

Phone: 0223 277314

JT Design Build Ltd
Bush House
72 Prince Street
BRISTOL BS1 4HU

Phone: 0272 290651   Fax: 0272 290946

Centre for Study for Environmental Change
Lancaster University
Fylde College
LANCASTER
Lancashire LA1 4YF

Phone: 0524 65201 X 2844   Fax: 0524 846339 DL

London Waste Regulation Authority
Hampton House
20 Albert Embankment
LONDON SE1 7TT

Phone: 071-587 3000   Fax: 071-587 5258

National Council of Building Material Producers
26 Store Street
LONDON WC1E 7BT

Phone: 071-323 3770   Fax: 071-323 0307

National Energy Foundation
Rockingham Drive
Linford Wood
MILTON KEYNES
MK14 6EG

Phone: 0908 672787   Fax: 0908 662296

National House-Building Council
Buildmark House
Chiltern Avenue
AMERSHAM
Buckinghamshire HP6 5AP

Phone: 0494 434477   Fax: 0494 728521

National Radiological Protection Board
Chilton
DIDCOT
Oxfordshire OX11 ORQ

Phone: 0235 831600   Fax: 0235 833891

Addresses of relevant organisations

National Rivers Authority
Rivers House
Waterside Drive
Aztec West
Almondsbury
BRISTOL BS12 4UD

Phone: 0454 624400   Fax: 0454 624409

New Homes Environmental Group
82 New Cavendish Street
LONDON
W1M 8AD

Phone: 071-580 5588

New Homes Marketing Board
82 New Cavendish Street
LONDON
W1M 8AD

Phone: 071-580 5588   Fax: 071-323 1697

RIBA (Royal Institute of British Architects)
66 Portland Place
LONDON WIN 4AD

Phone: 071-580 5533   Fax: 071-255 1541

Royal Institution of Chartered Surveyors
12 Great George Street
Parliament Square
LONDON SW1P 3AD

Phone: 071-222 7000   Fax: 071-222 9430

The Royal Town Planning Institute
26 Portland Place
LONDON W1N 4BE

Phone: 071-636 9107   Fax: 071-323 1582

Timber Research and Development Association
Stocking Lane
Hughenden Valley
HIGH WYCOMBE
Buckinghamshire HP14 4ND

Phone: 0494 563691   Fax: 0494 565487

Forests Forever Campaign
Timber Trade Federation
Clareville House
26/27 Oxendon Street
LONDON SW1Y 4EL

Phone: 071-839 1891   Fax: 071-930 0094

UK Ecolabelling Board
7th Floor
Eastbury House
30-34 Albert Embankment
LONDON SE1 7TL

Phone: 071-820 1199   Fax: 071-820 1104

Wood Panel Products Federation
1 Hanworth Road
FELTHAM
Middlesex TW13 5AF

Phone: 081-751 6107   Fax: 081-890 2870

World Wide Fund for Nature
Panda House
Weyside Park
GODALMING
Surrey GU7 1XR

Phone: 0483 426444   Fax: 0483 426409

Environmental Checklist: Design

Project: **Project Number:**

Name: **Date:**

    ☑ Date   Initials

Stage 1 Agreeing the brief and setting the project environmental policy

1.1 Gaining commitment ☐

1.2 Agreeing the design brief ☐

1.3 Setting environmental policy

    1.3.1 Consultants' potential influence on the environmental aspects of a project ☐

    1.3.2 Setting environmental policy and the possible role of environmental/management systems ☐

1.4 Briefing the design team and assigning responsibility ☐

Stage 2 Inception and feasibility

2.1 Legislation and policy ☐

2.2 Environmental assessment ☐

2.3 Site issues

    2.3.1 Derelict and contaminated land ☐

    2.3.2 Site hydrology and water quality ☐

    2.3.3 Flora, fauna and landscape ☐

    2.3.4 Landfill gas and naturally-occurring methane and CO_2 ☐

    2.3.5 Non-ionizing radiation ☐

    2.3.6 Radon and thoron ☐

    2.3.7 Noise ☐

    2.3.8 Services infrastructure ☐

    2.3.9 Local community ☐

    2.3.10 Local climate – assess impact of the local climate on development ☐☐

    2.3.11 Local transport infrastructure ☐

    2.3.12 Cultural features

2.4 Securing outline planning permission and identifying financial implications ☐

Stage 3 Primary design choices and scheme design

3.1 Briefing the design team ☐ ▭

3.2 Legislation and policy

    3.2.1 Planning legislation ☐ ▭

    3.2.2 Building Regulations ☐ ▭

    3.2.3 The Wildlife and Countryside Act 1981 ☐ ▭

3.3 Site layouts and environmental impacts

    3.3.1 Re-use of land ☐ ▭

    3.3.2 Ecological value of site ☐ ▭

    3.3.3 Transport and access to site ☐ ▭

    3.3.4 Local transport infrastructure ☐ ▭

    3.3.5 Building microclimate ☐ ▭

    3.3.6 Overshadowing and access to daylight and sunlight ☐ ▭

    3.3.7 Passive solar design and its effect on site layouts ☐ ▭

    3.3.8 Earth sheltered design ☐ ▭

3.4 Impacts on occupants and users, visitors and neighbours

    3.4.1 User and community consultation ☐ ▭

    3.4.2 External appearance (aesthetics) ☐ ▭

    3.4.3 Noise ☐ ▭

    3.4.4 Anticipating and minimising development impacts ☐ ▭

3.5 Detailed consultation with relevant bodies

    3.5.1 Water quality and economy ☐ ▭

    3.5.2 Floodwater provisions ☐ ▭

    3.5.3 Archaeological and historical issues ☐ ▭

3.6 Designing appropriate landscaping ☐ ▭

3.7 Energy options

    3.7.1 Energy conservation and efficiency ☐ ▭

    3.7.2 Use of high levels of insulation to minimise energy use in buildings ☐ ▭

    3.7.3 Passive solar design of domestic buildings ☐ ▭

    3.7.4 Renewable energy sources ☐ ▭

    3.7.5 Combined heat and power ☐ ▭

    3.7.6 Design of civil engineering works for minimal energy use ☐ ▭

3.8 Labelling and other environmental information

    3.8.1 Use of energy labelling schemes ☐ ▭

    3.8.2 Green labelling of materials and products ☐ ▭

    3.8.3 BREEAM ☐ ▭

3.9 Ventilation options in buildings

    3.9.1 Air infiltration and ventilation rates ☐

    3.9.2 Natural ventilation and passive stack ventilation in small buildings ☐

    3.9.3 Ventilation in large structures and buildings ☐

    3.9.4 Mechanical ventilation and heat recovery ☐

    3.9.5 Air conditioning ☐

    3.9.6 Legionnaires' Disease ☐

3.10 Daylight and need for artificial light ☐

3.11 Occupant comfort ☐

3.12 Criteria for primary material selection

    3.12.1 Material for the primary structure ☐

    3.12.2 Brick, block and masonry construction ☐

    3.12.3 Cladding materials ☐

    3.12.4 Roofing materials ☐

    3.12.5 Timber ☐

    3.12.6 Services and materials ☐

    3.12.7 Insulants ☐

3.13 Use of materials

    3.13.1 Waste, salvage, recycling and re-use of materials ☐

    3.13.2 Design to minimise use of materials ☐

    3.13.3 Storage of recyclable materials ☐

    3.13.4 Design for maintenance and cleaning ☐

Stage 4 Detailed design, working drawings and specifications

4.1 Ensuring the design team knows the project environmental policy ☐

4.2 Legislation and policy

    4.2.1 Building Regulations (As 3.2.2 but repeated for ease of use) ☐

    4.2.2 Avoidance of hazardous materials ☐

    4.2.3 Radon and thoron – legal issues ☐

    4.2.4 Radon, thoron and naturally-occurring methane – technical issues ☐

    4.2.5 Contaminated land and toxic substances ☐

    4.2.6 Electromagnetic radiation – legal issues ☐

    4.2.7 Electro magnetic radiation – technical issues ☐

    4.2.8 Safety, security and fire ☐

4.3 Use of energy

    4.3.1 Heating ☐

    4.3.2 Cooling ☐

    4.3.3 Insulation ☐

4.4 Labelling and other environmental information

    4.4.1 Energy labelling schemes ☐

    4.4.2 BREEAM ☐

4.5 Landscape, ecology and the use of plants

    4.5.1 Landscape and ecology ☐

    4.5.2 Use of plants to 'green' the built environment ☐

4.6 Internal environment

    4.6.1 Services ☐

    4.6.2 Water use and disposal ☐

    4.6.3 Noise and acoustics ☐

    4.6.4 Ventilation and air quality ☐

    4.6.5 Lighting ☐

    4.6.6 Condensation ☐

    4.6.7 Controls ☐

    4.6.8 Legionnaires' Disease ☐

    4.6.9 Sick Building Syndrome ☐

4.7 Materials

    4.7.1 Use of environmentally acceptable materials ☐

    4.7.2 Environmental policy on materials ☐

    4.7.3 Criteria for selection ☐

    4.7.4 Avoidance of CFCs, HCFCs and reduction in NO_x, CO_2 and SO_x ☐

    4.7.5 Sourcing of timber ☐

    4.7.6 Transport of materials and components ☐

    4.7.7 Vetting suppliers ☐

    4.7.8 Use of recycled materials ☐

4.8 End-use considerations

    4.8.1 Environmental issues in respect of end use ☐

    4.8.2 Operational and post-operational requirements ☐

    4.8.3 Commissioning ☐

4.9 Handover of project environmental policy to contractor, and/or other consultants ☐

Stage 5 Environmental considerations at tendering and contract letting ☐

Index

For main legislation, see Part B of the main references – page 142. Further cross references are listed in the text.

acoustics, 4.6.3, *see also* noise
aesthetics, 2.3.9, 2.3.12, 3.4.2
agreeing the assignment or design brief, 1.2
agreeing the brief and setting the project environmental
 policy, stage 1
air conditioning, 3.9.5, 4.3.2, 4.6.3, 4.6.4, 4.6.8,
 4.6.9, 4.8.3
air infiltration and ventilation, 3.9.1, 3.9.3, 3.9.4
air quality, 4.6.4, 4.6.9
archaeological and historical issues, 2.3.2, 3.4.2, 3.5.3
assignment or design brief, agreement of, 1.2
avoidance of CFCs, HCFCs and reduction in
 NO_x, CO_2 and SO_x, 4.7.4
avoidance of hazardous materials, 4.2.2

block construction, 3.12.2
BREEAM, 1.2, 2.2, 2.3.3, 3.3.6, 3.8.3, 3.13.3,
 4.4.1, 4.4.2, stage 5
BRECSU, 3.7.1, 3.8.1, 3.9.3, 4.4.1, 4.6.1, 4.6.5, 4.7.3
brick construction, 3.12.2, 4.7.3
briefing the design team, 1.2, 1.4, 3.1, 4.1
BS7750, 1.3.2
BSRIA Code of Practice, 1.4, 3.9.5, 3.9.6, 3.12.6, 4.3.2,
 4.6.1,
 4.6.2, 4.6.7, 4.8.1, 4.8.2, 4.8.3, 4.9
building microclimate, 3.3.5
Building Regulations 3.2.2 (repeated at 4.2.1)
built environment, greening of, 4.5.2

CFCs, avoidance of, 3.9.5, 4.7.4
civil engineering works, 3.3.4, 4.7.6
 design for minimal energy use, 3.7.6
cladding materials, 3.12.3
cleaning, design for, 3.13.4
climate,
 effect on potential development, 2.3.10
 micro-, 3.3.5, 4.5.2
CO_2, reduction in, 3.7.1, 4.7.4
combined heat and power, 3.7.5
comfort, 3.4.3, 3.10, 3.11, 4.5.4, 4.6.9
commissioning, 4.8.3
commitment, gaining, 1.1
community
 consultation, 2.3.9, 3.4.1, 3.4.3, 3.4.4
 local, 2.3.9, 2.3.11, 3.3.4, 4.3.4
condensation, 4.6.6
consultants' potential influence on environmental
 aspects of a project, 1.3.1
consultation,
 user and community, 3.4.1
 with relevant bodies, 3.4.1, 3.5
contaminated land, 2.3.1, 3.3.1, 4.2.5
contract letting, environmental considerations, stage 5
controls, 3.11, 4.6.7, 4.6.9, 4.8.2
cooling, 3.9, 3.11, 4.3.2
criteria for material selection, 3.12, 4.7.3
cultural features, 2.3.12, 3.4.2

daylight
 and need for artificial light, 3.3.6, 3.3.7, 3.10,
 4.6.5
 and sunlight, access to, 3.3.5, 3.3.6
derelict land, 2.3.1, 3.3.1

design
 agreement of brief, 1.2, 1.4, 3.1, 4.1
 checklist, page 146
 earth sheltered, 3.3.8
 for maintenance and cleaning, 3.13.4
 of appropriate landscaping, 3.6
 of civil engineering works for minimal energy use,
 3.7.6
 team briefing, 1.4, 3.1, 4.1
 to minimise use of materials, 3.13.2
detailed working drawings and specifications, stage 4
development impacts,
 anticipating and minimising, 2.3.9, 3.4.4

earth sheltered design, 3.3.8
ecological value of site, 3.3.2
ecology, 2.3.3, 4.5.1
electromagnetic radiation
 legal issues, 4.2.6
 technical issues, 4.2.7
end-use considerations, 4.8
Energy Efficiency Office, 3.7.1, 3.7.4, 3.8.1,
 3.9.3, 4.4.1,
energy
 conservation and efficiency, 3.7, 3.9.2, 3.9.4,
 4.3, 4.6.7, 4.7.3
 labelling schemes, 3.8.1, 4.4.1
 options, 3.7
 renewable, 3.7.3, 3.7.4, 3.7.6
 use of, 3.7.2, 3.7.4, 4.3
engineering works, civil 3.3.4, 3.7.6, 4.7.6
environmental
 assessment, 2.2, 3.2.1
 considerations at tendering and contract letting, stage 5
 issues in respect of end use, 4.8.1
 management systems, 1.3.2, 1.4
 policy, stage 1, 4.9, stage 5
 policy on materials, 4.7.2
 statement, 2.2
environmentally acceptable materials, 4.7.1
external appearance (aesthetics), 2.3.12, 3.4.2

fauna, 2.3.3
fire, 4.2.8, 4.8.3
floodwater provisions, 3.5.2
flora, fauna and landscape, 2.3.3

green labelling of materials and products, 3.8.2
greening of the built environment, 4.5.2

handover of project environmental policy to
 contractor, and/or other consultants, 4.9
hazardous materials, avoidance of, 4.2.2
HCFCs, avoidance of, 3.9.5, 4.7.4
heating, 3.3.7, 3.7.1, 4.3.1, 4.6.8
historic buildings, 2.3.12
historical issues, 2.3.2, 3.4.2, 3.5.3

identifying financial implications, 2.4
impacts of development, 2.3.12, 3.3.4, 3.4, 3.5.3, 3.6
inception and feasibility, stage 2
insulation, 3.3.8, 3.7.2, 3.12.7, 4.3.3
internal environment, 2.3.7, 3.4.3, 3.9, 4.3, 4.6

labelling and other environmental information, 3.8, 4.4
land, re-use, 2.3.1, 3.3.1
landfill gas and naturally-occurring methane
    and CO_2, 2.3.4
landscape, 2.3.3, 4.5.1
landscaping, 2.3.3, 3.4.2, 3.5.2, 3.6
Legionnaires' Disease, 3.9.5, 3.9.6, 4.6.8
legislation
    and policy, 2.1, 3.2, 4.2 *see also* Part B of the
        references, page 142
    on planning, 3.2.1
light, artificial, 3.10
lighting, 4.6.5
local issues
    climate – assess impact on development, 2.3.10
    community, 2.3.9, 2.3.11, 3.3.4, 3.4
    transport infrastructure, 2.3.11, 3.3.4

maintenance, design for, 3.13.4, 4.6.8, 4.8.2
masonry construction, 3.12.2
materials, 4.7
    and components, transport of, 4.7.6
    avoidance of hazardous, 4.2.2
    cladding, 3.12.3
    criteria for selection, 3.12, 4.7.3
    design to minimise use of, 3.13.2
    environmental policy on. 4.7.2
    environmentally acceptable, 4.7.1
    for roofing, 3.12.4
    for the primary structure, 3.12.1
    storage of recyclable, 3.13.3
    timber, 3.12.5, 4.7.5
    use of recycled, 4.7.8
    use of 3.13
    waste, recycling and re-use of, 3.13.1
mechanical ventilation and heat
    recovery, 3.9.4
methane, 2.3.1, 2.3.4
microclimate, 3.3.5, 4.5.2

natural ventilation and passive stack ventilation
    in small buildings, 3.9.2
neighbours, impact on, 2.3.11
noise, 2.3.7, 2.3.9, 3.4.3, 4.6.1, 4.6.9
NO_x, reduction in, 4.7.4

occupant comfort, 3.11
operational and post-operational requirements,
    stage 1, 4.8.2
overshadowing and access to daylight and
    sunlight, 3.3.6
outline planning permission, 2.4
orientation, of a building 3.3.6, 3.3.7

passive solar design, 3.3.7, 3.7.3, 4.3.2, 4.6.5
planning
    legislation, 3.2.1
    permission, 2.4, 3.2.1
plants, use, 2.3.3, 3.3.5, 3.3.6, 3.3.7, 3.4.3, 3.6, 3.7.3,
    4.5, 4.5.2, 4.6.3
pollution, 2.3.1, 2.3.4, 2.3.5, 2.3.6, 3.3.1, 3.4.3, 3.4.4,
    3.5.1, 3.5.3, 3.9.6, 3.13.1, 3.13.4, 4.2.2, 4.2.3, 4.2.4,
    4.2.5, 4.2.6, 4.2.7, 4.7.4, 4.8.2, 4.8.3
primary design choices and scheme design, stage 3
project environmental policy, stage 1, 4.9, stage 5

radiation,
    electromagnetic, 2.3.5, 4.2.6, 4.2.7
    non-ionizing, 2.3.5
radon and thoron, 2.3.6, 4.2.3, 4.2.4
re-use
    of land, 3.3.1
    of materials, 3.13.1
recyclable materials, storage of, 3.13.3
recycled materials, use of, 4.7.8
recycling, 3.13.1
renewable energy sources, 3.7.4
responsibility, 1.4, 4.9
role of environmental management systems, 1.3.2
roofing materials, 3.12.4

safety, security and fire, 4.2.8
securing outline planning permission and
    identifying financial implications, 2.4
security, 4.2.8
services, 2.3.8, 3.12.6, 4.6.1
setting environmental policy, 1.3, 1.3.2
Sick Building Syndrome, 3.11, 4.6.9
site
    access, 3.3.3
    ecological value, 3.3.2
    hydrology and water quality, 2.3.2
    issues, 2.3
    layouts and environmental impacts, 3.3
sites of special scientific interest, 2.3.3
solar design, 3.3.7, 3.7.3
sourcing of timber, 4.7.5, 4.7.7
SO_x, reduction in, 4.7.4
specifications, stage 4
SSSIs, 2.3.3
storage of recyclable materials, 3.13.3
suppliers, vetting, 4.7.7

tendering, environmental considerations at, stage 5
thoron, 2.3.6, 4.2.3, 4.2.4
timber, 3.12.5
    sourcing of, 4.7.5
transport
    and access to site, 2.3.11, 3.3.3, 3.3.4, 4.7.6
    of materials and components, 4.7.6

user and community consultation, 3.4.1

ventilation
    and air quality, 4.6.4, 4.6.8, 4.6.9
    in large structures and buildings, 3.9.3, 3.9.4,
        3.9.6, 4.3.2
    natural, 3.9.2, 4.6.9
    options in buildings, 3.9
    stack, 3.9.2
vetting suppliers, 4.7.7

waste, recycling and re-use of materials, 3.13.1, 4.7.8
water
    economy, 3.5.1
    quality, 3.5.1, 3.9.6
    use and disposal, 4.6.2
Wildlife and Countryside Act 1981, 3.2.3
working drawings, stage 4